T0256663

Network Congestion Control

WILEY SERIES IN COMMUNICATIONS NETWORKING & DISTRIBUTED SYSTEMS.

Series Editor: David Hutchison, *Lancaster University*
Series Advisers: Harmen van As, *TU Vienna*
                   Serge Fdida, *University of Paris*
                   Joe Sventek, *Agilent Laboratories, Edinburgh*

The 'Wiley Series in Communications Networking & Distributed Systems' is a series of expert-level, technically detailed books covering cutting-edge research and brand new developments in networking, middleware and software technologies for communications and distributed systems. The books will provide timely, accurate and reliable information about the state-of-the-art to researchers and development engineers in the Telecommunications and Computing sectors.

Other titles in the series:

Wright: *Voice over Packet Networks* 0-471-49516-6   (February 2001)
Jepsen: *Java for Telecommunications* 0-471-49826-2   (July 2001)
Sutton: *Secure Communications* 0-471-49904-8   (December 2001)
Stajano: *Security for Ubiquitous Computing* 0-470-84493-0   (February 2002)
Martin-Flatin: *Web-Based Management of IP Networks
                       and Systems*   0-471-48702-3   (September 2002)
Berman, Fox, Hey: *Grid Computing. Making the Global
                      Infrastructure a Reality*   0-470-85319-0   (March 2003)
Turner, Magill, Marples: *Service Provision. Technologies for
             Next Generation Communications*   0-470-85066-3   (April 2004)
Welzl: *Network Cogestion Control: Managing Internet Traffic*
                                      0-470-02528-X   (July 2005)
Heckmann: *The Competitive Internet Service Provider: Network
             Architecture, Interconnection, Traffic Engineering and
             Network Design*   0-470-01293-5   (April 2006)
Raz: *Fast and Efficient Context-Aware Services* 0-470-01668-X   (April 2006)

Network Congestion Control
Managing Internet Traffic

Michael Welzl

University of Innsbruck, Austria

John Wiley & Sons, Ltd

Copyright © 2005   John Wiley & Sons Ltd, The Atrium, Southern Gate, Chichester,
West Sussex PO19 8SQ, England

Telephone (+44) 1243 779777

Email (for orders and customer service enquiries): cs-books@wiley.co.uk
Visit our Home Page on www.wiley.com

Reprinted with corrections November 2006.

This publication is designed to provide accurate and authoritative information in regard to the subject matter covered. It is sold on the understanding that the Publisher is not engaged in rendering professional services. If professional advice or other expert assistance is required, the services of a competent professional should be sought.

Other Wiley Editorial Offices

John Wiley & Sons Inc., 111 River Street, Hoboken, NJ 07030, USA

Jossey-Bass, 989 Market Street, San Francisco, CA 94103-1741, USA

Wiley-VCH Verlag GmbH, Boschstr. 12, D-69469 Weinheim, Germany

John Wiley & Sons Australia Ltd, 42 McDougall Street, Milton, Queensland 4064, Australia

John Wiley & Sons (Asia) Pte Ltd, 2 Clementi Loop #02-01, Jin Xing Distripark, Singapore 129809

John Wiley & Sons Canada Ltd, 22 Worcester Road, Etobicoke, Ontario, Canada M9W 1L1

Wiley also publishes its books in a variety of electronic formats. Some content that appears in print may not be available in electronic books.

Library of Congress Cataloging-in-Publication Data

Welzl, Michael, 1973–
Network congestion control : managing Internet traffic / Michael Welzl.
    p. cm.
Includes bibliographical references and index.
ISBN-13: 978-0-470-02528-4 (cloth : alk. paper)
ISBN-10: 0-470-02528-X (cloth : alk. paper)
1. Internet. 2. Telecommunication–Traffic–Management. I. Title.
TK5105.875.I57W454 2005
004.67′8 – dc22

                         2005015429

British Library Cataloguing in Publication Data

A catalogue record for this book is available from the British Library

ISBN-13 978-0-470-02528-4
ISBN-10 0-470-02528-X

Typeset in 10/12pt Times by Laserwords Private Limited, Chennai, India

All my life, I enjoyed (and am still enjoying) a lot of support from many people – family, friends and colleagues alike, ranging from my grandmother and my girlfriend to my Ph.D. thesis supervisors. I sincerely thank them all for helping me along the way and dedicate this book to every one of them. This is not balderdash, I really mean it!

Contents

Foreword

The Internet is surely the second most extensive machine on the planet, after the public switched telephone network (PSTN), and it is rapidly becoming as ubiquitous. In fact, the distinction between the two is fast diminishing as the vision of the unified telecommunication network begins to be realized, and telecommunication operators deploy voice over IP (VoIP) technology. One of the biggest issues involved in the transition from PSTN to VoIP is ensuring that the customer sees (hears!) the best possible Quality of Service at all times. This is a considerable challenge for the network's designers and engineers.

Meanwhile, national governments – and also the European Commission – are implicitly assuming the emergence of the 'Information Society', and even funding research in pursuit of it. Critical applications including health, education, business and government are going to be increasingly dependent on information networks, which will inevitably be based on Internet (and Web) technologies. The penetration of broadband access into homes as well as businesses is rapidly bringing Web (and Internet) into everyone's lives and work.

The Internet was never foreseen as the more commercial network that it has now become: an informal tool for researchers has become a cornerstone of business. It is crucial that the underlying technology of the Internet is understood by those who plan to employ it to support critical applications. These 'enterprise owners', whether they be governments or companies, need to understand the principles of operation of the Internet, and along with these principles, its shortcomings and even its vulnerabilities. It does have potential shortcomings, principally its unproven ability to act as a critical support infrastructure; and it does have vulnerabilities, including its inability to cope with distributed denial-of-service attacks. These are arguably among the most pressing topics for Internet research.

It is particularly important that there are no unwarranted assumptions about the ability of the Internet to support more commercial activities and various critical applications. People involved in managing and operating Internet-based networks, and those who are considering its potential, will be suitably educated by Michael Welzl's book.

Congestion – the overloading of switches or routers with arriving traffic packets – is a consequence of the design of the Internet. Many mechanisms have been proposed to deal with it, though few have been deployed as yet. This book covers the theory and practical considerations of congestion, and gives an in-depth treatment of the subject.

'Network Congestion Control: Managing Internet Traffic' is a welcome addition to the Wiley Series in Communications Networking & Distributed Systems.

David Hutchison
Lancaster University
April 2005

Preface

Some people raised their eyebrows when I told them that I was writing a book on congestion control, and said, 'Is this topic really large enough for a book?' Well, it certainly is, and I am sure that it is needed. For example, there are quite a number of Ph.D. students out there who work in this field–yet, when they start out, they do not have a comprehensive introductory book about the subject. This book is for them and, of course, for anyone else with an interest in this fascinating topic–graduate students, teachers in academia and network administrators alike.

While the original page estimate was only a little lower than the actual outcome, I am now convinced that it would have been possible to write a book of twice this size on congestion control–but this would have meant diverging from the original goals and including things that are already nicely covered in other places. Instead of overloading this book, I therefore choose to recommend two books that were published last year as complementary material: (Hassan and Jain 2004) and (Srikant 2004).

Even if there is only one author, no book is the work of a single person. In my case, there are many people who provided help in one way or another – Anil Agarwal, Simon Bailey, Sven Hessler and Murtaza Yousaf proofread the text; this was sometimes a hectic task, especially towards the end of the process, but they all just kept on providing me with valuable input and constructive criticism. Neither you nor I would be happy with the result if it was not for these people. I would like to point out that I never personally met Anil–we got in touch via a technical discussion in the end-to-end interest mailing list of the IRTF, and he just volunteered to proofread my book. This certainly ranks high in the list of 'nicest things that ever happened to me', and deserves a big thanks.

I would like to thank Craig Partridge for providing me with information regarding the history of congestion control and allowing me to use his description of the 'global congestion collapse' incident. Further thanks go to Martin Zwicknagl for his Zillertaler Bauernkrapfen example, Stefan Hainzer for bringing an interesting article about fairness to my attention, and Stefan Podlipnig for numerous discussions which helped to shape the book into its present form. Two tools are described in Appendix A, where it is stated that they 'were developed at our University'. Actually, they were implemented by the following students under my supervision: Christian Sternagel wrote CAVT, and Wolfgang Gassler, Robert Binna and Thomas Gatterer wrote NSBM. The congestion control behaviour analyses of various applications described in Chapter 6 were carried out by Muhlis Akdag, Thomas Rammer, Roland Wallnöfer, Andreas Radinger and Marcus Fischer under my supervision. I would like to mention Michael Trawöger because he insisted that he be named *here*; he had a 'cool cover idea' that may or may not have made it onto the final book's front page. While the right people to thank for this are normally the members of the Wiley graphics department, if the cover of this book is the best that you have ever seen, it surely was his influence.

I would like to thank two people whom I have never been in touch with but the work that they have done had such a major influence on this book: Raj Jain and Grenville Armitage. Like nobody else, Raj Jain is able to explain seemingly complex things in a simple manner; reading his papers assured me that it would indeed be possible to write a book that is easily comprehensible yet covers the whole range of congestion control issues. I had the writing style of his early papers in mind when I wrote certain sections of this book, especially Chapter 2. Grenville Armitage had a somewhat similar influence, as he impressed me with his book 'Quality of Service in IP Networks' (Armitage 2000). It is easily readable and introductory, but it still manages to cover all the important topics. I used this book as some sort of a 'role model'—Grenville Armitage basically did with QoS what I tried to do with congestion control.

I should not forget to thank the people who helped me at the publisher's side of the table—David Hutchison, Birgit Gruber, Joanna Tootill and Julie Ward. These are of course not the only people at John Wiley & Sons who were involved in the production of this book—while I do not know the names of the others, I thank them all! Figures 3.13, 4.11, 6.1 and A.5 were taken from (Welzl 2003) with kind permission of Springer Science and Business Media (Kluwer Academic Publishers at the time the permission was granted).

Finally, I would like to name the people whom I already mentioned in the 'dedication': my Ph.D. thesis supervisors, who really did a lot for me, were Max Mühlhäuser and Jon Crowcroft. My grandmother, Gertrud Welzl, provided me with an immense amount of support throughout my life, and I also enjoy a lot of support from my girlfriend, Petra Ratzenböck; I really strained her patience during the final stages of book writing. As I write this, it strikes me as odd to thank my girlfriend, while most other authors thank their wives—perhaps the time has come to change this situation.

List of tables

List of figures

1

Introduction

Congestion control is a topic that has been dealt with for a long time, and it has also become a facet of daily life for Internet users. Most of us know the effect: downloading, say, a movie trailer can take five minutes today and ten minutes tomorrow. When it takes ten minutes, we say that the network is congested. Those of us who have attended a basic networking course or read a general networking book know some things about how congestion comes about and how it is resolved in the Internet – but this is often just the tip of the iceberg.

On the other hand, we have researchers who spend many years of their lives with computer networks. These are the people who read research papers, take the time to study the underlying math, and write papers of their own. Some of them develop protocols and services and contribute to standardization bodies; congestion control is their daily bread and butter. But what about the people in between – those of us who would like to know a little more about congestion control without having to read complicated research papers, and those of us who are in the process of *becoming* researchers?

Interestingly, there seems to be no comprehensive and easily readable book on the market that fills this gap. While some general introductory networking books do have quite detailed and well-written parts on congestion control – a notable example is (Kurose and Ross 2004) – it is clearly an important and broad enough topic to deserve an introductory book of its own.

1.1   Who should read this book?

This book is the result of an attempt to describe a seemingly complex domain in simple words. In the literature, all kinds of methods are applied to solve problems in congestion control, often depending on the background of authors – from fuzzy logic to game theory and from control theory to utility functions and linear programming, it seems that quite a diverse range of mathematical tools can be applied. In order to understand all of these papers, one needs to have a thorough understanding of the underlying theory. This may be a little too much for someone who would just like to become acquainted with the field

Network Congestion Control: Managing Internet Traffic   Michael Welzl
© 2005 John Wiley & Sons, Ltd

(e.g. a network administrator who is merely interested in some specifics about the dynamic behaviour of network traffic).

In my opinion, starting with these research papers is also inefficient. It is a waste of time for Ph.D. students, who typically should finalize a thesis within three or four years; rather, what they would need at the very beginning of their endeavour is a book that gives them an overview without becoming too involved in the more sophisticated mathematical aspects. As an example, consider a Ph.D. student who has to develop a new mechanism that builds upon the notion that Internet users should be cooperative for their own good (as with most common peer-to-peer file sharing tools). In some introductory papers, she might read about how different controls influence fairness – which might lead her to become lost in the depths of control theory, whereas a game-theoretic viewpoint could have pointed to an easy solution of the problem.

One could argue that learning some details about control theory is not the worst idea for somebody who wants to become involved in congestion control. I agree, but this is also a question of time – one can only learn so many things in a day, and getting on the right track fast is arguably desirable. This is where this book can help: it could be used as a roadmap for the land of congestion control. The Ph.D. student in our example could read it, go 'hey, game theory is what I need!' and then proceed with the bibliography. This way, she is on the right track from the beginning.

By providing an easily comprehensible overview of congestion control issues and principles, this book can also help graduate students to broaden their knowledge of how the Internet works. Usually, students attain a very rough idea of this during their first networking course; follow-up courses are often held, which add some in-depth information. Together with other books on special topics such as 'Routing in the Internet' (Huitema 2000) and 'Quality of Service in IP Networks' (Armitage 2000), this book could form the basis for such a specialized course. To summarize, this book is written for:

Ph.D. students who need to get on track at the beginning of their thesis.

Graduate students who need to broaden their knowledge of how the Internet works.

Teachers who develop a follow-up networking course on special topics.

Network administrators who are interested in details about the dynamic behaviour of network traffic.

1.2 Contents

In computer networks literature, there is often a tendency to present *what* exists and *how* it works. The intention behind this book, on the other hand, is to explain *why* things work the way they do. It begins with an explanation of fundamental issues that will be helpful for understanding the design rationales of the existing and envisioned mechanisms, which are explained afterwards. The focus is on principles; here are some of the things that you will not find in it:

Mathematical models: While the ideas behind some mathematical models are explained in Chapter 2, going deeply into such things would just have complicated the book and

would have shifted it away from the fundamental goal of being easy to read.
Recommended alternative: Rayadurgam Srikant, 'The Mathematics of Internet Congestion Control', Springer Verlag 2004 (Srikant 2004).

Performance evaluations: You will not find results of simulations or real-life measurements that show that mechanism X performs better than mechanism Y. There are several reasons for this: first, it is not the intention to prove that some things work better than others – it is not even intended to judge the quality of mechanisms here. Rather, the goal is to show what has been developed, and why things were designed the way they are. Second, such results often depend on aspects of X and Y that are not relevant for the explanation, but they would have to be explained in order to make it clear why X and Y behave the way they do. This would only distract the reader, and it would therefore also deviate from the original goal of being an easily comprehensible introduction. Third, the performance of practically every mechanism that is presented in this book was evaluated in the paper where it was first described, and this paper can be found in the bibliography.
Recommended alternative: Mahbub Hassan and Raj Jain, 'High Performance TCP/IP Networking', Pearson Education International 2004 (Hassan and Jain 2004).

Exhaustive descriptions: Since the focus is on principles (and their application), you will not find complete coverage of each and every detail of, say, TCP (which nevertheless makes up quite a part of this book). This is to say that there are, for example, no descriptions of 'tcpdump' traces.
Recommended alternative: W. Richard Stevens, 'TCP/IP Illustrated, Volume 1: The Protocols', Addison-Wesley Publishing Company 1994 (Stevens 1994).

Since this book is of an introductory nature, it is not necessary to have an immense amount of background knowledge for reading it; in particular, one does not have to be a mathematics genius in order to understand even the more complicated parts, as equations were avoided wherever possible. It is however assumed that the reader knows some general networking fundamentals, such as

- the distinction between connection oriented and connectionless communication;

- what network layers are and why we have them;

- how basic Internet mechanisms like HTTP requests and routing roughly work;

- how checksums work and what 'Forward Error Correction' (FEC) is all about;

- the meaning of terms such as 'bandwidth', 'latency' and 'end-to-end delay'.

All these things can be learned from general introductory books about computer networks, such as (Kurose and Ross 2004), (Tanenbaum 2003) and (Peterson and Davie 2003), and they are also often covered in a first university course on networking. A thorough introduction to concepts of performance is given in (Sterbenz et al. 2001).

1.3 Structure

While this book is mostly about the Internet, congestion control applies to all packet-oriented networks. Therefore, Chapter 2 is written in a somewhat general manner and explains the underlying principles in a broad way even though they were mainly applied to (or brought up in the context of) Internet protocols. This book does not simply say 'TCP works like this' – rather, it says 'mechanism a has this underlying reasoning and works as follows' in Chapter 2 and 'this is how TCP uses mechanism a' in Chapter 3.

In this book, there is a clear distinction between things that are standardized and deployed as opposed to things that should be regarded as research efforts. Chapter 3 presents technology that you can expect to encounter in the Internet of today. It consists of two parts: first, congestion control in end systems is explained. In the present Internet, this is synonymous with the word 'TCP'. The second part focuses on congestion control – related mechanisms within the network. Currently, there is not much going on here, and therefore, this part is short: we have an active queue management mechanism called 'RED', and we may still have the ATM 'Available Bit Rate (ABR)' service operational in some places. The latter is worth looking at because of its highly sophisticated structure, but its explanation will be kept short because the importance (and deployment) of ATM ABR is rapidly declining.

Chapter 4 goes into details about research endeavours that may or may not become widely deployed in the future. Some of them *are* already deployed in some places (for example, there are mechanisms that transparently enhance the performance of TCP without requiring any changes to the standard), but they have not gone through the IETF procedure for specification and should probably not be regarded as parts of the TCP/IP standard. Topics include enhancements that make TCP more robust against adverse network effects such as link noise, mechanisms that perform better than TCP in high-speed networks, mechanisms that are a better fit for real-time multimedia applications, and RED improvements. Throughout this chapter, there is a focus on practical, rather than theoretical works, which either have a certain chance of becoming widely deployed one day or are well known enough to be regarded as representatives for a certain approach.

The book is all about efficient use of network capacities; on a longer time scale, this is 'traffic management'. While traffic management is not the main focus of this book, it is included because issues of congestion control and traffic management are indeed related. The main differences are that traffic management occurs on a longer time scale, often relies on human intervention, and control is typically executed in a different place (not at connection endpoints, which are the most commonly involved elements for congestion control). Traffic management tools typically fall into one of two categories: 'traffic engineering', which is a means to influence routing, and 'Quality of Service' (QoS) – the idea of providing users with differentiated and appropriately priced network services. Both these topics are covered in Chapter 5, but this part of the book is very brief in order not to stray too far from the main subject. After all, while traffic engineering and QoS are related, they simply do not fall in the 'congestion control' category.

Chapter 6 is specifically written for researchers (Ph.D. students in particular) who are looking for ideas to work on. It is quite different from anything else in the book: while the goal of the rest is to inform the reader about specific technology and its underlying ideas and principles, the intention of this chapter is to show that things are still far from perfect in practice and to point out potential research avenues. As such, this chapter is also extremely biased – it could be seen as a collection of my own thoughts and views about the future of congestion control. You may agree with some of them and completely disagree with others; like a good technical discussion, going through such potentially controversial material should be thought provoking rather than informative. Ideally, you would read this chapter and perhaps even disagree with my views but you would be stimulated to come up with better ideas of your own.

The book ends with two appendices: first, the problem of teaching congestion control is discussed. Personally, I found it quite hard to come up with practical congestion control exercises that a large number of students can individually solve within a week. There appeared to be an inevitable trade-off between exposure to the underlying dynamics (the 'look and feel' of things) on the one hand and restraining the additional effort for learning how to use certain things (which does not relate to the problem itself) on the other. As a solution that turned out to work really well, two small and simple Java tools were developed. These applications are explained in Appendix A, and they are available from the accompanying website of this book, `http://www.welzl.at/congestion`.

Appendix B provides an overview of related IETF work. The IETF, the standardization body of the Internet, plays a major role in the area of congestion control; its decisions have a large influence on the architecture of the TCP/IP stacks in the operating systems of our home PCs and mechanisms that are implemented in routers alike. Historically, Internet congestion control has also evolved from work in the IETF, and quite a large number of the citations in the bibliography of this book are taken from there. Note that this appendix does not contain a thorough description of the standardization process – rather, it is a roadmap to the things that have been written.

1.3.1 Reader's guide

It is probably every author's dream that readers would go through the book from the beginning to the end, without ever losing attention in the fascinating material that is presented. In reality, this is quite rare, and it may be better to assume that most people will only use a book to look up certain details or read some chapters or sections that are relevant for them. If you are one of them, this section is for you – it is a list of what to read, depending on the type of reader you are:

The interested reader without a strong background in networks
should read Chapters 2 and 3, and perhaps also Chapter 5.

The knowledgeable reader who is only interested in research efforts
should browse Chapters 4 and 6.

The hurried reader should read the specific parts of choice (e.g. if the goal is to gain an understanding of TCP, Chapter 3 should be read), use Chapters 2 and 5 only to look up information and avoid Chapter 6, which does not provide any essential congestion control information.

Appendix A is for teachers, and Appendix B is for anybody who is not well acquainted with the IETF and wants to find related information fast.

I tried hard to make this book not only informative but also an enjoyable read. I know that this is not easily achieved, and a joke here and there does not really make a book more pleasant to use. For me, the *structure* of a book largely dictates whether I enjoy working with it or not – if I am in a hurry, I do certainly *not* enjoy reading poorly organized books – thus, this is something I tried hard to avoid. Equipped with the information from this introductory chapter, Appendix B, the index and bibliography of this book, you should be able to efficiently use it and not waste your time. Have fun!

2

Congestion control principles

2.1 What is congestion?

Unless you are a very special privileged user, the Internet provides you with a service that is called *best effort*; this means that the network simply does its best to deliver your data as efficiently as possible. There are no guarantees: if I do my best to become a movie star, I might actually succeed – but then again, I might not (some people will tell you that you *will* succeed if you just keep trying, but that is a different story). The same is true of the packets that carry your data across the Internet: they might reach the other end very quickly, they might reach somewhat slower or they might never even make it. Downloading a file today could take twice as long as it took yesterday; a streaming movie that had intolerable quality fluctuations last night could look fine tomorrow morning. Most of us are used to this behaviour – but where does it come from?

There are several reasons, especially when the timescale we are looking at is as long as in these examples: when Internet links become unavailable, paths are recalculated and packets traverse different inner network nodes ('routers'). It is well known that even the weather may have an influence if a wireless link is involved (actually, a friend of mine who accesses the Internet via a wireless connection frequently complains about bandwidth problems that seem to correspond with rainfall); another reason is – you guessed it – *congestion*.

Congestion occurs when resource demands exceed the capacity. As users come and go, so do the packets they send; Internet performance is therefore largely governed by these inevitable natural fluctuations. Consider an ISP that would allow up to 1000 simultaneous data flows (customers), each of which would have a maximum rate of 1 Mbps but an average rate of only 300 kbps. Would it make sense to connect their Internet gateway to a 1 Gbps link (which means that all of them could be accommodated at all times), or would, say, 600 Mbps be enough? For simplicity, let us assume that the ISP chooses the 600 Mbps option for now because this link is cheaper and suffices most of the time.

In this case, the gateway would see occasional traffic spikes that go beyond the capacity limit as a certain number of customers use their maximum rate at the same time. Since these excess packets cannot be transferred across the link, there are only two things that this device can do: buffer the packets or drop them. Since such traffic spikes are typically

Network Congestion Control: Managing Internet Traffic   Michael Welzl
© 2005 John Wiley & Sons, Ltd

limited in time, standard Internet routers usually place excess packets in a buffer, which roughly works like a basic FIFO ('First In, First Out') queue and only drop packets if the queue is full. The underlying assumption of this design is that a subsequent traffic reduction would eventually drain the queue, thus making it an ample device to compensate for short traffic bursts. Also, it would seem that reserving enough buffer for a long queue is a good choice because it increases the chance of accommodating traffic spikes. There are however two basic problems with this:

1. Storing packets in a queue adds significant delay, depending on the length of the queue.

2. Internet traffic does not strictly follow a Poisson distribution, that is, the assumption that there are as many upward fluctuations as there are downward fluctuations may be wrong.

The consequence of the first problem is that packet loss can occur no matter how long the maximum queue is; moreover, because of the second problem, queues should generally be kept short, which makes it clear that not even defining the upper limit is a trivial task. Let me repeat this important point here before we continue:

> *Queues should generally be kept short.*

When queues grow, the network is said to be congested; this effect will manifest itself in increasing delay and, at worst, packet loss.

Now that we know the origin and some of the technical implications of congestion, let us find a way to describe it. There is no 'official', universally accepted definition of network congestion; this being said, the most elaborate attempt was probably made in (Keshav 1991a). Here is a slightly simplified form of this definition, which acknowledges that the truly important aspect of network performance is not some technical parameter but user experience:

> *A network is said to be congested from the perspective of a user if the service quality noticed by the user decreases because of an increase in network load.*

2.1.1 Overprovisioning or control?

Nowadays, the common choice of ISPs is to serve the aforementioned 1000 flows with 1 Gbps or even more in order to avoid congestion within their network. This method is called *overprovisioning*, or, more jovially, 'throwing bandwidth at the problem'. The Internet has made a transition from a state of core overload to a state of core underload; congestion has, in general, moved into the access links. The reasons for this are of a purely financial nature:

• Bandwidth has become cheap. It pays off to overprovision a network if the excess bandwidth costs significantly less than the amount of money that an ISP could expect to lose in case a customer complains.

• It is more difficult to control a network that has just enough bandwidth than an overprovisioned one. Network administrators will require more time to do their task and perhaps need special training, which means that these networks cost more money. Moreover, there is an increased risk of network failures, which once again leads to customer complaints.

- With an overprovisioned network, an ISP is prepared for the future – there is some headroom that allows the accommodation of an increasing number of customers with increasing bandwidth demands for a while.

The goal of *congestion control* mechanisms is simply to use the network as efficiently as possible, that is, attain the highest possible throughput while maintaining a low loss ratio and small delay. Congestion must be *avoided* because it leads to queue growth and queue growth leads to delay and loss; therefore, the term 'congestion avoidance' is sometimes used. In today's mostly uncongested networks, the goal remains the same – but while it appears that existing congestion control methods have amply dealt with overloaded links in the Internet over the years, the problem has now shifted from 'How can we get rid of congestion?' to 'How can we make use of all this bandwidth?'. Most efforts revolve around the latter issue these days; while researchers are still pursuing the same goal of efficient network usage, it has become somewhat fashionable to replace 'congestion control' with terms such as 'high performance networking', 'high speed communication' and so on over the last couple of years. Do not let this confuse you – it is the same goal with slightly different environment conditions. This is a very important point, as it explains why we need congestion control at all nowadays. Here it is again:

> *Congestion control is about using the network as efficiently as possible. These days, networks are often overprovisioned, and the underlying question has shifted from 'how to eliminate congestion' to 'how to efficiently use all the available capacity'. Efficiently using the network means answering both these questions at the same time; this is what good congestion control mechanisms do.*

The statement 'these days, networks are often overprovisioned' appears to imply that it has not always been this way. As a matter of fact, it has not, and things may even change in the future. The authors of (Crowcroft et al. 2003) describe how the ratio of core to access bandwidth has changed over time; roughly, they state that excess capacity shifts from the core to access links within 10 years and swings back over the next 10 years, leading to repetitive 20-year cycles. As an example, access speeds were higher than the core capacity in the late 1970s, which changed in the 1980s, when ISDN (56 kbps) technology came about and the core was often based upon a 2 Mbps Frame Relay network. The 1990s were the days of ATM, with 622 Mbps, but this was also the time of more and more 100 Mbps Ethernet connections.

As mentioned before, we are typically facing a massively overprovisioned core nowadays (thanks to optical networks which are built upon technologies such as *Dense Wavelength Division Multiplexing (DWDM)*), but the growing success of Gigabit and, more recently, 10 Gigabit Ethernet as well as other novel high-bandwidth access technologies (e.g. UMTS) seems to point out that we are already moving towards a change. Whether it will come or not, the underlying mechanisms of the Internet should be (and, in fact, *are*) prepared for such an event; while 10 years may seem to be a long time for the telecommunications economy, this is not the case for TCP/IP technology, which has already managed to survive several decades and should clearly remain operational as a binding element for the years to come.

On a side note, moving congestion to the access link does not mean that it will vanish; if the network is used in a careless manner, queues can still grow, and increased delay and packet loss can still occur. One reason why most ISPs see an uncongested core these days is that the network is, in fact, *not* used carelessly by the majority of end nodes – and when it is, these events often make the news ('A virus/worm has struck again!'). An amply provisioned network that can cope with such scenarios may not be affordable. Moreover, as we will see in the next section, the heterogeneity of link speeds along an end-to-end path that traverses several ISP boundaries can also be a source of congestion.

2.2   Congestion collapse

The Internet first experienced a problem called *congestion collapse* in the 1980s. Here is a recollection of the event by Craig Partridge, Research Director for the Internet Research Department at BBN Technologies (Reproduced by permission of Craig Partridge):

> Bits of the network would fade in and out, but usually only for TCP. You could ping. You could get a UDP packet through. Telnet and FTP would fail after a while. And it depended on where you were going (some hosts were just fine, others flaky) and time of day (I did a lot of work on weekends in the late 1980s and the network was wonderfully free then). Around 1pm was bad (I was on the East Coast of the US and you could tell when those pesky folks on the West Coast decided to start work...).
>
> Another experience was that things broke in unexpected ways – we spent a lot of time making sure applications were bullet-proof against failures. One case I remember is that lots of folks decided the idea of having two distinct DNS primary servers for their subdomain was silly – so they'd make one primary and have the other one do zone transfers regularly. Well, in periods of congestion, sometimes the zone transfers would repeatedly fail – and voila, a primary server would timeout the zone file (but know it was primary and thus start authoritatively rejecting names in the domain as unknown).
>
> Finally, I remember being startled when Van Jacobson first described how truly awful network performance was in parts of the Berkeley campus. It was far worse than I was generally seeing. In some sense, I felt we were lucky that the really bad stuff hit just where Van was there to see it.[1]

One of the earliest documents that mention the term 'congestion collapse' is (Nagle 1984) by John Nagle; here, it is described as a stable condition of degraded performance that stems from unnecessary packet retransmissions. Nowadays, it is, however, more common to refer to 'congestion collapse' when a condition occurs where *increasing* sender rates *reduces* the total throughput of a network. The existence of such a condition was already acknowledged in (Gerla and Kleinrock 1980) (which even uses the word 'collapse' once to describe the behaviour of a throughput curve) and probably earlier – but how does it arise?

---

[1]Author's note: Van Jacobson brought congestion control to the Internet; a significant portion of this book is based upon his work.

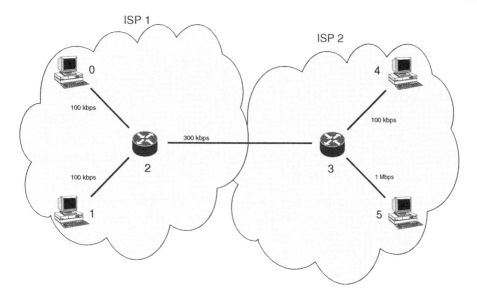

Figure 2.1 Congestion collapse scenario

Consider the following example: Figure 2.1 shows two service providers (ISP 1 and ISP 2) with two customers each; they are interconnected with a 300 kbps link[2] and do not know each other's network configuration. Customer 0 sends data to customer 4, while customer 1 sends data to customer 5, and both sources always send as much as possible (100 kbps); there is no congestion control in place. Quite obviously, ISP 1 will notice that its outgoing link is not fully utilized (2 * 100 kbps is only 2/3 of the link capacity); thus, a decision is made to upgrade one of the links. The link from customer 0 to the access router (router number 2) is upgraded to 1 Mbps (giving customers too much bandwidth cannot hurt, can it?). At this point, you may already notice that it would have been a better decision to upgrade the link to router 2 because the link that connects the corresponding sink to router 3 has a higher capacity – but this is unknown to ISP 1.

Figure 2.2 shows the throughput that the receivers (customers 4 and 5) will see before (a) and after (b) the link upgrade. These results were obtained with the 'ns' network simulator[3] (see A.2): each source started with a rate of 64 kbps and increased it by 3 kbps every second. In the original scenario, throughput increases until both senders reach the capacity limit of their access links. This result is not surprising – but what happens when the bandwidth of the 0–2 link is increased? The throughput at 4 remains the same because it is always limited to 100 kbps by the connection between nodes 3 and 4. For the connection from 1 to 5, however, things are a little different. It goes up to 100 kbps (its maximum rate – it is still constrained to this limit by the link that connects customer 1 to router 2); as the rate approaches the capacity limit, the throughput curve becomes smoother

---

[2]If you think that this number is unrealistic, feel free to multiply all the link bandwidth values in this example with a constant factor x – the effect remains the same.

[3]The simulation script is available from the accompanying web page of the book, http://www.welzl.at/congestion

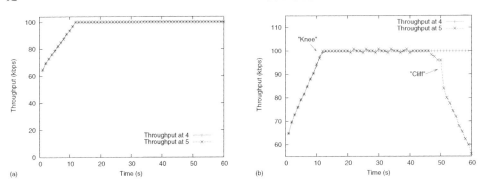

Figure 2.2   Throughput before (a) and after (b) upgrading the access links

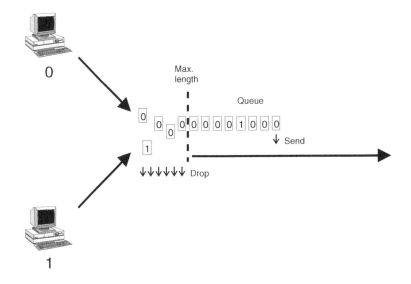

Figure 2.3   Data flow in node 2

(this is called the *knee*), and beyond a certain point, it suddenly drops (the so-called cliff) and then decreases further.

The explanation for this strange phenomenon is congestion: since both sources keep increasing their rates no matter what the capacities beyond their access links are, there will be congestion at node 2 – a queue will grow, and this queue will have more packets that stem from customer 0. This is shown in Figure 2.3; roughly, for every packet from customer 1, there are 10 packets from customer 0. Basically, this means that the packets from customer 0 unnecessarily occupy bandwidth of the bottleneck link that could be used by the data flow (just 'flow' from now on) coming from customer 1 – the rate will be narrowed down to 100 kbps at the 3–4 link anyway. The more the customer 0 sends, the greater this problem.

If customer 0 knew that it would never attain more throughput than 100 kbps and would therefore refrain from increasing the rate beyond this point, customer 1 could stay at its limit of 100 kbps. A technical solution is required for appropriately reducing the rate of customer 0; this is what congestion control is all about. In (Jain and Ramakrishnan 1988), the term 'congestion control' is distinguished from the term 'congestion avoidance' via its operational range (as seen in Figure 2.2 (b)): schemes that allow the network to operate at the knee are called *congestion avoidance schemes*, whereas congestion control just tries to keep the network to the left of the cliff. In practice, it is hard to differentiate mechanisms like this as they all share the common goal of maximizing network throughput while keeping queues short. Throughout this book, the two terms will therefore be used synonymously.

2.3 Controlling congestion: design considerations

How could one design a mechanism that automatically and ideally tunes the rate of the flow from customer 0 in our example? In order to find an answer to this question, we should take a closer look at the elements involved:

- Traffic originates from a *sender*; this is where the first decisions are made (when to send how many packets). For simplicity, we assume that there is only a single sender at this point.

- Depending on the specific network scenario, each packet usually traverses a certain number of *intermediate nodes*. These nodes typically have a queue that grows in the presence of congestion; packets are dropped when it exceeds a limit.

- Eventually, traffic reaches a *receiver*. This is where the final (and most relevant) performance is seen – the ultimate goal of almost any network communication code is to maximize the satisfaction of a user at this network node. Once again, we assume that there is only one receiver at this point, in order to keep things simple.

Traffic can be controlled at the sender and at the intermediate nodes; performance measurements can be taken by intermediate nodes and by the receiver. Let us call members of the first group *controllers* and members of the second group *measuring points*. Then, at least one controller and one measuring point must participate in any congestion control scheme that involves feedback.

2.3.1 Closed-loop versus open-loop control

In control theoretic terms, systems that use feedback are called *closed-loop* control as opposed to *open-loop* control systems, which have no feedback. Systems with nothing but open-loop control have some value in real life; as an example, consider a light switch that will automatically turn off the light after one minute. On the other hand, neglecting feedback is clearly not a good choice when it comes to dissolving network congestion, where the dynamics of the system – the presence or absence of other flows – dictate the ideal behaviour.

In a computer network, applying open-loop control would mean using *a priori* knowledge about the network – for example, the bottleneck bandwidth (Sterbenz et al. 2001). Since, as explained at the beginning of this chapter, the access link is typically the bottleneck nowadays, this property is in fact often known to the end user. Therefore, applications that ask us for our network link bandwidth during the installation process or allow us to adjust this value in the system preferences probably apply perfectly reasonable open-loop congestion control (one may hope that this is not all they do to avoid congestion). A network that is solely based on open-loop control would use resource reservation, that is, a new flow would only enter if the *admission control* entity allows it to do so. As a matter of fact, this is how congestion has always been dealt with in the traditional telephone network: when a user wants to call somebody but the network is overloaded, the call is simply rejected. Historically speaking, admission control in connection-oriented networks could therefore be regarded as a predecessor of congestion control in packet networks.

Things are relatively simple in the telephone network: a call is assumed to have fixed bandwidth requirements, and so the link capacity can be divided by a pre-defined value in order to calculate the number of calls that can be admitted. In a multi-service network like the Internet however, where a diverse range of different applications should be supported, neither bandwidth requirements nor application behaviour may be known in advance. Thus, in order to efficiently utilize the available resources, it might be necessary for the admission control entity to measure the actual bandwidth usage, thereby adding feedback to the control and deviating from its strictly open character. Open-loop control was called *proactive* (as opposed to *reactive* control) in (Keshav 1991a). Keshav also pointed out what we have just seen: that these two control modes are not mutually exclusive.

2.3.2   Congestion control and flow control

Since intermediate nodes can act as controllers and measuring points at the same time, a congestion control scheme could theoretically exist where neither the sender nor the receiver is involved. This is, however, not a practical choice as most network technologies are designed to operate in a wide range of environment conditions, including the smallest possible setup: a sender and a receiver, interconnected via a single link. While congestion collapse is less of a problem in this scenario, the receiver should still have some means to slow down the sender if it is busy doing more pressing things than receiving network packets or if it is simply not fast enough. In this case, the function of informing the sender to reduce its rate is normally called *flow control*.

The goal of flow control is to protect the receiver from overload, whereas the goal of congestion control is to protect the network. The two functions lend themselves to combined implementations because the underlying mechanism is similar: feedback is used to tune the rate of a flow. Since it may be reasonable to protect both the receiver and the network from overload at the same time, such implementations should be such that the sender uses a rate that is the minimum of the results obtained with flow control and congestion control calculations. Owing to these resemblances, the terms 'flow control' and 'congestion control' are sometimes used synonymously, or one is regarded as a special case of the other (Jain and Ramakrishnan 1988).

2.4   Implicit feedback

Now that we know that a general-purpose congestion control scheme will normally have the sender tune its rate on the basis of feedback from the receiver, it remains to be seen whether control and/or measurement actions from within the network should be included. Since it seems obvious that adding these functions will complicate things significantly, we postpone such considerations and start with the simpler case of *implicit* feedback, that is, measurements that are taken at the receiver and can be used to deduce what happens within the network.

In order to determine what such feedback can look like, we must ask the question, What can happen to a packet as it travels from source to destination? From an end-node perspective, there are basically three possibilities:

1. It can be delayed.

2. It can be dropped.

3. It can be changed.

Delay can have several reasons: distance (sending a signal to a satellite and back again takes longer than sending it across an undersea cable), queuing, processing in the involved nodes, or retransmissions at the link layer. Similarly, packets can be dropped because a queue length is exceeded, a user is not admitted, equipment malfunctions, or link noise causes a checksum of relevance to intermediate systems to fail. Changing a packet could mean altering its header or its content (payload). If the content changed but the service provided by the end-to-end protocol includes assurance of data integrity, the data carried by the packet become useless, and the conclusion to be made is that some link technology in between introduced errors (and no intermediate node dropped the packet due to a checksum failure). Such errors usually stem from link noise, but they may also be caused by malicious users or broken equipment. If the header changed, we have some form of explicit communication between end nodes and inner network nodes – but at this point, we just decided to ignore such behaviour for the sake of simplicity. We do not regard the inevitable function of placing packets in a queue and dropping them if it overflows as such active participation in a congestion control scheme.

The good news is that the word 'queue' was mentioned twice at the beginning of the last paragraph – at least the factors 'delay' and 'packet dropped' can indicate congestion. The bad news is that each of the three things that can happen to a packet can have quite a variety of reasons, depending on the specific usage scenario. Relying on these factors therefore means that implicit assumptions are made about the network (e.g. assuming that increased delay always indicates queue growth could mean that it is assumed that a series of packets will be routed along the same path). They should be used with care.

Note that we do not have to restrict our observations to a single packet only: there are quite a number of possibilities to deduce network properties from end-to-end performance measurements of series of packets. The so-called *packet pair* approach is a prominent example (Keshav 1991a). With this method, two packets are sent back-to-back: a large packet immediately followed by a small packet. Since it is reasonable to assume that there is a high chance for these packets to be serviced one after another at the bottleneck, the interspacing of these packets can be used to derive the capacity of the bottleneck link. While this method clearly makes several assumptions about the behaviour of routers along the path, it yields a metric that could be valuable for a congestion control mechanism (Keshav 1991b). For the sake of simplicity, we do not discuss such schemes further at this point and reserve additional observations for later (Section 4.6.3).

2.5   Source behaviour with binary feedback

Now that we have narrowed down our considerations to implicit feedback only, let us once again focus on the simplest case: a notification that tells the source 'there was congestion'. Packet loss is the implicit feedback that could be interpreted in such a manner, provided that packets are mainly dropped when queues overflow; this kind of feedback was used (and this assumption was made) when congestion control was introduced in the Internet. As you may have already guessed, the growing use of wireless (and therefore noisy) Internet connections poses a problem because it leads to a misinterpretation of packet loss; we will discuss this issue in greater detail later.

What can a sender do in response to a notification that simply informs it that the network is congested? Obviously, in order to avoid congestion collapse, it should reduce its rate. Since it does not make much sense to start with a fixed rate and only reduce it in a network where users could come and go at any time, it would also be useful to find a rule that allows the sender to increase the rate when the situation within the network has enhanced. The relevant information in this case would therefore be 'there was no congestion' – a message from the receiver in response to a packet that *was* received. So, we end up with a sender that keeps sending, a receiver that keeps submitting binary yes/no feedback, and a rule for the sender that says 'increase the rate if the receiver says that there was no congestion, decrease otherwise'. What we have not discussed yet is *how* to increase or decrease the rate.

Let us stick with the simple congestion collapse scenario depicted in Figure 2.1 – two senders, two receivers, a single bottleneck link – and assume that both flows operate in a strictly synchronous fashion, that is, the senders receive feedback and update their rate at the same time. The goal of our rate control rules is to efficiently use the available capacity, that is, let the system operate at the 'knee', thereby reducing queue growth and loss. This state should obviously be reached as soon as possible, and it is also clear that we want the system to maintain this state and avoid oscillations. Another goal that we have not yet taken into consideration is fairness – clearly, if all link capacities were equal in Figure 2.1, we would not want one user to fully utilize the available bandwidth while the other user obtains nothing. Fairness is in fact a somewhat more complex issue, which we will further examine towards the end of this chapter; for now, it suffices to stay with our simple model.

2.5.1   MIMD, AIAD, AIMD and MIAD

If the rate of a sender at time t is denoted by $x(t)$, $y(t)$ represents the binary feedback with values 0 meaning 'no congestion' and 1 meaning 'congestion' and we restrict our observations to linear controls, the rate update function can be expressed as

$$x(t+1) = \begin{cases} a_i + b_i x(t) & \text{if } y(t) = 0 \\ a_d + b_d x(t) & \text{if } y(t) = 1 \end{cases} \tag{2.1}$$

where a_i, b_i, a_d and b_d are constants (Chiu and Jain 1989). This linear control has both an additive (a) and a multiplicative component (b); if we allow the influence of only one component at a time, this leaves us with the following possibilities:

- $a_i = 0$; $a_d = 0$; $b_i > 1$; $0 < b_d < 1$
  Multiplicative Increase, Multiplicative Decrease (MIMD)

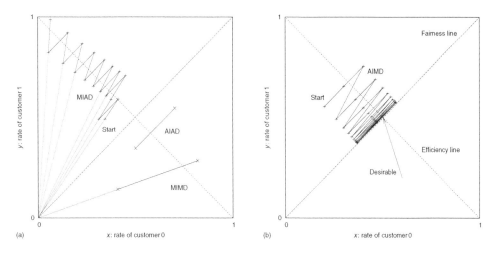

Figure 2.4 Vector diagrams showing trajectories of AIAD, MIMD, MIAD (a) and AIMD (b)

- $a_i > 0$; $a_d < 0$; $b_i = 1$; $b_d = 1$
  Additive Increase, Additive Decrease (AIAD)

- $a_i > 0$; $a_d = 0$; $b_i = 1$; $0 < b_d < 1$
  Additive Increase, Multiplicative Decrease (AIMD)

- $a_i = 0$; $a_d < 0$; $b_i > 1$; $b_d = 1$
  Multiplicative Increase, Additive Decrease (MIAD)

While these are by no means all the possible controls as we have restricted our observations to quite a simple case, it may be worth asking which ones out of these four are a good choice.

The system state transitions given by these controls can be regarded as a trajectory through an n-dimensional vector space – in the case of two controls (which represent two synchronous users in a computer network), this vector space is two dimensional and can be drawn and analysed easily. Figure 2.4 shows two vector diagrams with the four controls as above. Each axis in the diagrams represents a customer in our network. Therefore, any point (x, y) represents a two-user allocation. The sum of the system load must not exceed a certain limit, which is represented by the 'Efficiency line'; the load is equal for all points on lines that are parallel to this line. One goal of the distributed control is to bring the system as close as possible to this line.

Additionally, the system load consumed by customer 0 should be equal to the load consumed by customer 1. This is true for all points on the 'Fairness line' (note that the fairness is equal for all points on all lines that pass through the origin. Following (Chiu and Jain 1989), we therefore call any such line 'Equi-fairness line'). The optimal point is the point of intersection of the efficiency line and the fairness line. The 'Desirable' arrow in Figure 2.4 (b) represents the optimal control: it quickly moves to the optimal point and stays there (is stable). It is easy to see that this control is unrealistic for binary feedback: provided that both flows obtain the same feedback at any time, there is no way for one flow to interpret the information 'there is congestion' or 'there is no congestion' differently than the other – but the 'Desirable' vector has a negative x component and a positive y component. This means that the two flows make a different control decision at the same time.

Adding a constant positive or negative factor to a value at the same time corresponds to moving along at a 45° angle. This effect is produced by AIAD: both flows start at a point underneath the efficiency line and move upwards at an angle of 45°. The system ends up in an overloaded state (the state transition vector passes the efficiency line), which means that it now sends the feedback 'there is congestion' to the sources. Next, both customers decrease their load by a constant factor, moving back along the same line. With AIAD, there is no way for the system to leave this line.

The same is true for MIMD, but here, a multiplication by a constant factor corresponds with moving along an equi-fairness line. By moving upwards along an equi-fairness line and downwards at an angle of 45°, MIAD converges towards a totally unfair rate allocation, the customer in favour being the one who already had the greater rate at the beginning. AIMD actually approaches perfect fairness and efficiency, but because of the binary nature of the feedback, the system can only converge to an equilibrium instead of a stable point – it will eventually fluctuate around the optimum. MIAD and AIMD are also depicted in the 'traditional' (time = x-axis, rate = y-axis) manner in Figure 2.5 – these diagrams clearly show how the gap between the two lines grows in case of MIAD, which means that fairness is degraded, and shrinks in case of AIMD, which means that the allocation becomes fair.

The vector diagrams in Figure 2.4 (which show trajectories that were created with the 'Congestion Avoidance Visualization Tool' (*CAVTool*) – see Section A.1 for further details) are a simple means to illustrate the dynamic behaviour of a congestion control scheme. However, since they can only show how the rates evolve from a single starting point, they cannot be seen as a means to prove that a control behaves in a certain manner. In (Chiu and Jain 1989), an algebraic proof can be found, which states that the linear decrease policy should be multiplicative, and the linear increase policy should always have an additive component, and optionally may have a multiplicative component with the coefficient no less than one if the control is to converge to efficiency and fairness in a distributed manner.

Note that these are by no means all the possible controls: the rate update function could also be nonlinear, and we should not forget that we restricted our observations to

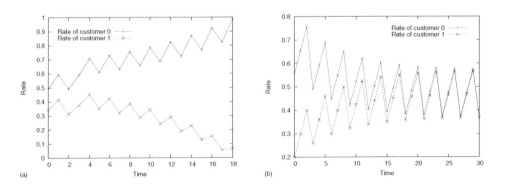

Figure 2.5 Rate evolvement with MIAD (a) and AIMD (b)

implicit binary feedback, which is not necessarily all the information that is available. Many variations have been proposed over the years; however, to this day, the source rate control rule that is implemented in the Internet basically is an AIMD variant, and its design can be traced back to the reasoning in this section.

2.6 Stability

It is worth taking another look at MIAD: as shown for an example in Figure 2.4, this mechanism converges to unfairness with a bias towards the customer that had a greater rate in the beginning. What if there is no such customer, that is, the trajectory starts at the fairness line? Since moving upwards along a line that passes through the origin and moving downwards at an angle of 45° means that the trajectory will never leave the fairness line, this control will eventually fluctuate around the optimum just like AIMD. The fairness of MIAD is, however, *unstable*: a slight deviation from the optimum will lead the control away from this point. This is critical because our ultimate goal is to use the mechanism in a rather uncontrolled environment, where users come and go at will. What if one customer would simply decide to stop sending for a while? All of a sudden, MIAD would leave the fairness line and allow the other customer to fully use the available capacity, leaving nothing for the first customer.

It is therefore clear that any control that is designed for use in a real environment (perhaps even with human intervention) should be *stable*, that is, it should not exhibit the behaviour of MIAD. This fact is true for all kinds of technical systems; as an example, we certainly do not want the autopilot of an aeroplane to abandon a landing procedure just because of strong winds. Issues of control and stability are much broader in scope than our area of interest (congestion control in computer networks). In engineering and mathematics, *control theory* generally deals with the behaviour of dynamic systems – systems that can be described with a set of functions (rules, equations) that specify how variables change over time. In this context, *stability* means that for any bounded input over any amount of time the output will also be bounded.

2.6.1 Control theoretic modelling

Figure 2.6 shows a simple closed-loop (or *feedback*) control loop. Its behaviour depends upon the difference between a reference value r and the output y of the system, the error e. The controller C takes this value as its input and uses it to change the inputs u to P, the system under control. A standard example for a real-life system that can be modelled with such a feedback control loop is a shower: when I slowly turn up the hot water tap from a starting point r, I execute control (my hand is C) and thereby change the input u (a certain amount of hot/cold water flowing through the pipes) to the system under control P (the water in the shower). The output y is the temperature – I feel it and use it to adjust the tap again (this is the feedback to the control).

This example comes in handy for explaining an important point of control theory: a controller should only be fed a system state that reflects its output. In other words, if I keep turning up the tap bit by bit and do not wait until the water temperature reaches a new level and stays there, I might end up turning up the hot water too quickly and burn myself (impatient as I am, this actually happens to me once in a while). This also applies to

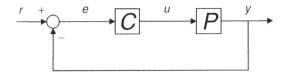

Figure 2.6 Simple feedback control loop

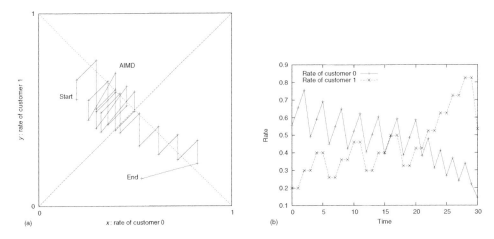

Figure 2.7 AIMD trajectory with $RTT_{\text{customer 0}} = 2 \times RTT_{\text{customer 1}}$

congestion control – a (low-pass) filter function should be used in order to pass only states to the controller that are expected to last long enough for its action to be meaningful (Jain and Ramakrishnan 1988). Action should be carried out whenever such feedback arrives, as it is a fundamental principle of control that the control frequency should be equal to the feedback frequency. Reacting faster leads to oscillations and instability while reacting slower makes the system tardy (Jain 1990).

2.6.2   Heterogeneous RTTs

The simple case of *n* synchronous flows sharing a single resource can be modelled with a feedback loop like the one in Figure 2.6; then, with such a model and the mathematical tools provided by control theory, the stability of the controller can be proven. Doing this is worthwhile: if it turns out that a mechanism is instable in this rather simple case, it is certainly useless. However, the opposite assumption is not valid because the scenario is too simplistic. In reality, senders hardly update their rate at exactly the same time – rather, there is usually an arbitrary number of asynchronously operating control loops that influence one another.

Figure 2.7 illustrates that even the stability of AIMD is questionable when control loops are not in sync. Here, the Round-trip Time (RTT) of customer 0 was chosen to be twice as long as the RTT of customer 1, which means that for every rate update of customer 0 there are two updates of customer 1. Convergence to fairness does not seem to occur with this example trajectory, and modelling it is mathematically sophisticated, potentially leading to somewhat unrealistic assumptions. For example, it is common to consider a 'fluid-flow model', where packets have a theoretical size of one bit and the rate of a data stream is therefore arbitrarily scalable; in practice, a packet of, say, 1000 bytes is either received as a whole or it is unusable. Often, network researchers rely on simulations for a deeper study of such scenarios. This being said, some authors have taken on the challenge of mathematically analysing stability of network congestion control in both the synchronous and the asynchronous case – two notable works in this area are (Johari and Tan 2001) and (Massoulie 2002); (Luenberger 1979) is a recommendable general introduction to dynamic systems and the notion of stability.

2.6.3 The conservation of packets principle

The seminal work that introduced congestion control to the Internet was 'Congestion Avoidance and Control', published by Van Jacobson at the ACM SIGCOMM 1988 conference (Jacobson 1988). In this paper, he suggested the execution of congestion control at the sources via a change to the 'Transmission Control Protocol' (TCP); as in our model, feedback is binary and implicit – packet loss is detected via a timeout and interpreted as congestion, and the control law is (roughly) AIMD. How did Van Jacobson take care of stability? The following quotes from his work shed some light on this matter:

> The flow on a TCP connection (..) should obey a 'conservation of packets' principle.

> By 'conservation of packets', we mean that for a connection 'in equilibrium', i.e. running stably with a full window of data in transit, the packet flow is what a physicist would call 'conservative': A new packet isn't put into the network until an old packet leaves. The physics of flow predicts that systems with this property should be robust in the face of congestion.

> A conservative flow means that for any given time, the integral of the packet density around the sender-receiver-sender loop is a constant. Since packets have to 'diffuse' around this loop, the integral is sufficiently continuous to be a Lyapunov function for the system. A constant function trivially meets the conditions for Lyapunov stability so the system is stable and any superposition of such systems is stable.

Two factors are crucial for this scheme to work:

1. *Window-based* (as opposed to *rate-based*) control

2. precise knowledge of the RTT.

2.7 Rate-based versus window-based control

There are two basic methods to throttle the rate of a sender (for simplicity, we assume only a single sender and receiver and no explicit help from intermediate routers): rate based and window based. Both methods have their advantages and disadvantages. They work as follows:

Rate-based control means that a sender is aware of a specific data rate (bits per second), and the receiver or a router informs the sender of a new rate that it must not exceed.

Window-based control has the sender keep track of a so-called *window* – a certain number of packets or bytes that it is allowed to send before new feedback arrives. With each packet sent, the window is decreased until it reaches 0. As an example, if the window is 6, the unit is packets, and no feedback arrives, the sender is allowed to send exactly six packets; then it must stop. The receiver accepts and counts incoming packets and informs the sender that it is allowed to increase the window by a certain amount. Since the sender's behaviour is very strictly dictated by the presence or absence of incoming feedback, window-based control is said to be *self-clocking*.

Rate-based control is simpler, and it is said to be more suitable for streaming media applications because it does not stop if no feedback arrives. In general, we want such applications to keep on sending no matter what happens as the data source often does the same. If, say, audio from a radio station is transmitted across the Internet, the moderator will normally not stop talking if the network is congested. Window-based control can show a certain stop-and-go behaviour, which is unwanted for such applications.

There is also an advantage to window-based control: it is a better match for the conservation of packets principle, as the 'you may now increase your window because I received a packet' feedback from the receiver is semantically equivalent to what the principle says: *a new packet is not put into the network until an old packet leaves*. From a network perspective, window-based flow control is perhaps generally less harmful because the sender will automatically stop sending when there is a massive problem in the network and no more feedback arrives.

A disadvantage of window-based control is that it can lead to traffic bursts. Consider Figure 2.8 (a), which shows the simple example case of a sender sending a full initial window of six (unacknowledged) packets and then waiting for feedback. As of yet, no packets have reached the receiver. While the sender transmitted the packets at a regular spacing (i.e. exactly one packet every x seconds), three of the six packets are enqueued at the bottleneck due to congestion in the network. Immediately after the snapshot shown in the figure, congestion is resolved and the three packets are sent on with regular spacing that depends on the bottleneck service rate.

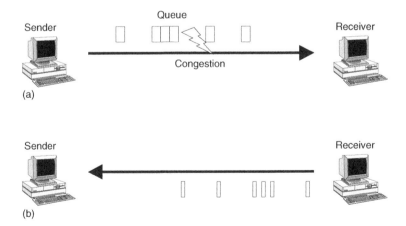

Figure 2.8 (a) The full window of six packets is sent. (b) The receiver ACKs

If this rate is higher than the sender rate (i.e. the flow does not fully saturate the capacity) and the receiver immediately sends an acknowledgement (ACK) upon reception of a packet, the network will look as shown in Figure 2.8 (b) after a while: the reduced spacing between the three packets that were enqueued is reflected by the spacing of ACKs. If the sender immediately increases its window by 1 when an ACK arrives and always sends packets as soon as it is allowed to do so, the reduced spacing between the three ACKs will lead to reduced spacing between the corresponding three data packets, and so forth. This effect also occurs when the ACKs (and not the data packets) experience congestion.

What we have here is a data burst, and there is no way to alleviate this effect unless either the sender or the receiver deliberately delays its reaction; this is called *pacing* (Sterbenz et al. 2001). On a side note, the reduced spacing between packets that were enqueued at the bottleneck is exploited by the packet pair approach (see Sections 2.4 and 4.6.3) in order to deduce information about the network.

Figure 2.8 shows another problem in addition to the effect of congestion: the window is too small. It is clearly undesirable to have the sender send six packets, then stop, wait for a while and then transmit the next six packets as the series of ACKs arrives. Rather, the sender should fully saturate the link, that is, also transmit packets during the second phase that is shown in Figure 2.8 (b). As a matter of fact, there should be no 'first' and 'second' phase – unless a problem occurs, packets should be transmitted continuously and ACKs should just keep arriving all the time.

In order to reach such a state, the sender must be able to increase its rate – hence, simply increasing the window by one packet in response to an ACK is not enough. Increasing the rate means to have the window grow by more than one packet per ACK, and decreasing it means reducing the window size. The ideal window size (which has the sender saturate the link) in bytes is the product of the bottleneck capacity and the RTT. Thus, in addition to the necessity of precise RTT estimation for the sake of self-clocking (i.e. adherence to the conservation of packets principle), the RTT can also be valuable for determining the ideal maximum window.

2.8 RTT estimation

The RTT is an important component of various functions:

- If reliable transmission is desired, a sender must retransmit packets if they are dropped somewhere in the network. The common way to realize this is to number packets consecutively and have the receiver acknowledge each of them; if an ACK is missing for a long time, the sender must assume that the packet was dropped and retransmit it. This mechanism, which is called *Automatic Repeat Request (ARQ)*, normally requires a timer that is initialized with a certain timeout value when a packet is sent. If this value is too large, it can take an unnecessarily long time until a packet is retransmitted – but if it is too small, a spurious retransmission could be caused (a packet that reached the receiver is sent a second time), thereby wasting network capacity and perhaps causing errors at the receiver side. Neglecting delay fluctuations from queuing and other interfering factors, the ideal timeout value seems to be one RTT, or at least a function thereof.

- Finding the right timeout value is also important in the context of congestion control with implicit binary packet loss feedback, that is, when packet loss is interpreted as a sign of congestion. Here, detecting loss unnecessarily late (because of a large timeout value) can cause harm: clearly, in the face of congestion, sources should reduce their rates as soon as possible. Once again, a timeout value that is too small is also a disadvantage, as it can lead to spurious congestion detection and therefore cause an unnecessary rate reduction, and the ideal value is most probably a function of an RTT.

- As we have seen in the previous section, the 'conservation of packets' principle mandates that in equilibrium a new packet is only sent into the network when an old packet leaves. This can only be realized if the sender has an idea of when packets leave the network. In a reliable protocol based on ARQ, they do so when the receiver generates the corresponding ACK; thus, the sender can be sure that it may send a new packet when it receives an ACK. Since the time from sending a packet to receiving the corresponding ACK is an RTT, this value has to play a role when changing the rate (which is in fact a bit more sophisticated than 'send and wait for ACK' based transmission – this mode, which is called *Stop-and-wait ARQ* does not allow to increase the rate beyond one acknowledged data unit (typically a packet) per RTT).

Measuring the duration between sending a packet and receiving the corresponding ACK yields the time it took for the packet to reach the receiver and for the ACK to come back; it is the most recent RTT measurement. Normally, the RTT of interest is in the future: the system is controlled on the basis of a state that it is assumed to have when the next packets will be sent. Since the RTT is dictated by things that happen within the network (delay in queues, path changes and so on), it depends on the state of the system and is not necessarily equal to the most recent measurement – solely relying on this value is a too simplistic approach. Rather, a prediction must be made using the history of RTT samples. This is not necessarily a simple process, as it should ideally reflect the environment (range of variations in queues, etc.). It is also essential to ensure that an ACK yields a true RTT and not, say, the time interval between the first transmission of a packet that was later dropped and the ACK that belongs to its retransmitted copy (Karn and Partridge 1995).

As a common rule of thumb, RTT prediction should be conservative: generally, it can be said that overestimating the RTT causes less harm than underestimating it. An RTT estimator should be fairly robust against short dips while ensuring appropriate reaction to significant peaks.

2.9   Traffic phase effects

In real networks, where there is no fluid-flow model but there are only fixed packets that either reach a node as a whole or fail completely (this is the essence of the so-called 'store and forward' switching deployed in most computer networks today), and where no two flows can transmit their packets across a single FIFO queue at exactly the same time, some ugly phenomena occur. Let us take a closer look at some of them – a study of these problems will show us that even RTT estimation and self-clocking have their downsides.

As a start, consider Figure 2.9 (a). Here, the throughput of three *Constant Bit Rate (CBR)* flows is shown. The diagram was obtained by simulation: three sources (nodes 0, 1 and 2) transmitted 1000 byte packets to three sinks (nodes 5, 6 and 7, respectively) across a single bottleneck. The topology is depicted in Figure 2.10; all links had a capacity of 1 Mbps and a link delay of 1 s. Note that the intention was not to model a realistic scenario

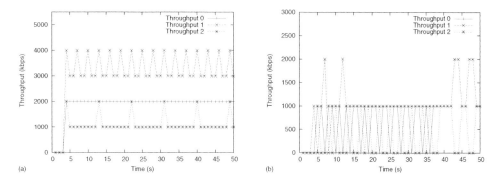

Figure 2.9 Three CBR flows – separate (a) and interacting (b)

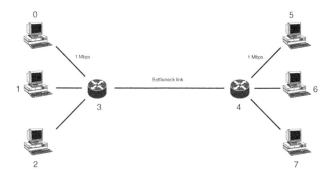

Figure 2.10 Topology used to simulate burstiness with CBR flows

(otherwise, 1 s would probably be too much) but merely to illustrate an effect. Source 0 sent a packet to destination 5 every 500 ms, source 1 sent a packet to destination 6 every 300 ms and source 2 sent a packet to destination 7 every 900 ms, which corresponds with rate values of 16, 26.6 and 8.8 kbps, respectively. Throughput was measured at the receiving nodes every second.

Since the network is clearly overprovisioned and all link bandwidths are equal, there is no congestion; therefore, one might expect all the lines in Figure 2.9 to be constant, but this is only true for the line corresponding with source 1. What this diagram shows is that, in reality, even a CBR flow does not exhibit constant behaviour when the sampling interval is not a multiple of its rate (as in the case of flow 1). At first sight, this may merely seem a bit surprising when creating plots, but it is, in fact, a serious issue that plays a significant role when any kind of control is executed at time instances that do not appropriately relate to the control intervals of the involved flows.

A FIFO queue is one such example. In order to make this control come into play, the same simulation as before was carried out, but this time, the bottleneck link capacity was set to 16 kbps – clearly, this is not enough to accommodate all flows at the same time. The throughput plots from this simulation (Figure 2.9 (b)) show that the impact of the FIFO queue distorted the output even more; the bandwidth is by no means fairly or proportionally divided among the flows. Flow 1, for instance, was unable to transmit a single packet to the receiver after 38 s. It seems that of the 170 packets that were dropped in this simulation quite a significant fraction belonged to this flow – but it is neither the flow with the highest nor the one with the lowest rate. Apparently, flow 1 just had the bad luck of transmitting packets at the wrong time instances (i.e. when the queue was full). This problem is called a *phase effect* (Floyd and Jacobson 1991).

2.9.1 Phase effects in daily life

The following real-life example may help you grasp this problem: in some places in Tyrol, when there is a reason to celebrate, the famous 'Zillertaler Bauernkrapfen' (a local dish) are sold. They only taste good when they are hot, and they are not easy to prepare (Drewes and Haid 2000); thus, they are not sold all the time. Instead, a dozen become available, are sold right away, and then it takes a while until the next dozen are ready. Now let us assume that you are eager for such a dish. You stand in line, wait and it turns out that the dozen are sold before it is your turn. Since it takes quite a while before the next dozen are ready and simply standing there means that you miss the rest of the celebration, you leave the booth and decide to return in half an hour.

Nobody told you that Bauernkrapfen become available exactly every 30 minutes, and so you end up being among the last people in the line, and once again the dozen are sold before you can have yours. This phase effect could occur this way forever unless you communicate with the controlling entity (the cook at the booth) or change your strategy. One possibility would be to add some randomness and return *approximately* after 30 minutes. Similarly, the cook could alleviate the problem by waiting for a certain random time interval before beginning to cook Zillertaler Bauernkrapfen again. As we will see in the next section, the latter strategy (adding randomness to servers) was chosen in order to solve this problem in the Internet (Floyd and Jacobson 1993); interestingly, the other variant (adding randomness to sources) was also suggested in order to further improve the situation (Diederich et al. 2000).

Note that congestion controlled flows which react on the basis of the RTT are by no means a better match for this problem than are CBR flows – in fact, things may look even worse: if, for example, a routing change suddenly causes 10 flows that roughly have the same RTT to transmit their packets across a saturated link at the same time, and the queue at the bottleneck can only hold nine packets, one of these 10 flows will experience a packet drop. This will lead to a sudden rate reduction, thereby making room for the other nine flows until each of them loses a packet, which means that there is now some headroom for the single flow to increase its rate again. Chances are that the flows will remain synchronized like this, and the single flow that experienced packet loss first will eventually obtain a significantly different network performance than each of the nine other flows. This effect is also called *global synchronization*.

2.10 Queue management

As we have seen, traffic phase effects occur when different flows see different performances. In our previous example, the fact that only nine out of ten flows could fit their first packet in the queue could be solved by simply increasing the buffer size in the router. It seems that these effects would not occur or could at least be significantly diminished by increasing the maximum queue length. Since a queue is only meant to compensate for sudden traffic bursts, one may wonder what would happen if the queue length was endless. Of course, there is no such thing as an endless buffer, but it could be *quite* long. Could such a buffer, together with well-chosen link capacities, prevent packet loss altogether?

2.10.1 Choosing the right queue length

As you may expect, the short and simple answer is 'no'. There are two reasons for this: first, the source behaviour that we have so far taken into consideration relies on packet loss as a congestion indicator – thus, the rate of sources will keep increasing until the queue length grows beyond its limit, no matter how high that limit is. Second, a queue can always overflow because of the very nature of network traffic, which usually shows at least some degree of self-similarity. Without going into further details at this point, we can explain this effect by looking at rainfall, which shares this property: since there is no guarantee for an equal number of sunny days and rainy days, you can never have a guarantee that a dam is large enough. Personally, I believe this to be the reason why we keep hearing about floods that occur in areas that are already known to be endangered.

There is another reason why just picking a very large number for the maximum queue length is not a good idea: queuing delay is a significant portion in the overall end-to-end delay, which should be as small as possible for obvious reasons (just consider telephony – delay is quite bothersome to users in this application). Remember what I said on page 8: *Queues should generally be kept short.* The added delay from queues also negatively influences a congestion control algorithm, which should obtain feedback that reflects the current state in the network and should not lag behind in time. As we have seen in Section 2.8, estimation of the RTT plays a major role for proper source behaviour – long queues distort the RTT samples and render any RTT-based mechanism inefficient.

After this discussion, we still do not know what the ideal maximum queue length is; it turns out that the proper tuning of this parameter is indeed a tricky issue. Let us look at a single flow and a single link for a moment. In order to perfectly saturate the link, it must have $c \times d$ bits in transit, where c is the capacity (in bits per second) and d is the delay of the link (in seconds). Thus, from an end-system performance perspective, links are best characterized by their *bandwidth* $\times$ *delay product*.[4] On the basis of this fact and the nature of congestion control algorithms deployed in the Internet, a common rule of thumb says that the queue limit of a router should be set to the bandwidth $\times$ delay product, where 'bandwidth' is the link capacity and 'delay' is the average RTT of flows that traverse it. Recently, it has been shown that this rule, which leads to quite a large buffer space in common Internet backbone routers (e.g. with the common average RTT choice of 250 ms, a 10 Gbps router requires a buffer space of 2.5 Gbits), is actually outdated, and that it would in fact be better to divide the bandwidth $\times$ delay product by the square root of the number of flows in the network (Appenzeller et al. 2004).

2.10.2 Active queue management

However, even if we use these research results to ideally tune the maximum queue length, the phase effect from the previous section will not vanish because control is still executed

---

[4]This is related to our finding in Section 2.7 that the ideal window size is the bandwidth $\times$ RTT product.

independent of the individual RTTs of flows; in other words, relying on only one such metric, the average RTT of all flows in the network, does not suffice. What needs to be done? As already mentioned in the previous section, introducing randomness in one of the controlling entities is a possible solution. In the case of the Internet, the chosen entity was the router; (Floyd and Jacobson 1993) describes a mechanism called *Random Early Detection (RED)*, which is now widely deployed and makes a decision to drop a packet on the basis of the average queue length and a random function as well as some parameters that are somewhat hard to tune. RED is a popular example of a class of so-called *active queue management* (AQM) mechanisms (Braden et al. 1998). In addition to alleviating traffic phase effects, a scheme like RED has the advantage of generally keeping the queue size (and hence end-to-end delay) low while allowing occasional bursts of packets in the queue.

What makes the design of such a scheme a difficult task is the range of possibilities to choose from: packets can be dropped from the front or the end of the queue, or they can be picked from the middle (which is usually inefficient because it leads to time-consuming memory management operations). There is a large variety of possible methods to monitor the queue size and use it in combination with some randomness to make a decision – functions that are applied in this context usually have their advantages as well as corresponding disadvantages. Perhaps the most-important design goal is *scalability*: if a mechanism that works perfectly with, say, ten flows, but ceases to work in the presence of thousands of flows because router resources do not suffice any longer in such a scenario (e.g. the available memory is exceeded), it cannot be used as a core element of a network that is as large as the Internet.

2.11   Scalability

Systems can scale in several dimensions – depending on the context, 'scalability' could mean that something works with a small or large amount of traffic, or that it will not cease to work if link capacities grow. In the context of computer networks, the most common is related to the number of users, or communication flows, in the system. If something scales, it is expected to work no matter how many users there are. Quite obviously, Internet technology turned out to scale very well, as is illustrated by the continuous growth of the network itself. Therefore, the Internet community has become quite religious about ensuring scalability at all times – and it appears that they are doing the right thing.

2.11.1   The end-to-end argument

When we are talking about the Internet, the key element of scalability is most certainly the *end-to-end argument*. This rule (or rather set of arguments with a common theme) was originally described in (Saltzer et al. 1984). It is often quoted to say that one should move complexities 'out of the network' (towards endpoints, upwards in the protocol stack) and keep the network itself as 'simple as possible'. This interpretation is actually incorrect because it is a little more restrictive than it should be. Still, it is reasonable to consider it as a first hint: if a system is designed in such a manner, the argument is clearly not violated. This is a good thing because strict adherence to the end-to-end argument is regarded as the primary reason for the immense scalability – and thus, success – of the Internet (Carpenter 1996). Here is the original wording from (Saltzer et al. 1984):

The function in question can completely and correctly be implemented only with the knowledge and help of the application standing at the end points of the communication system. Therefore, providing that questioned function as a feature of the communication system itself is not possible. (Sometimes an incomplete version of the function provided by the communication system may be useful as a performance enhancement.)

The difference between the original end-to-end argument and their stricter interpretation is that the argument is focused on application requirements. In other words, while application-specific functions should not be placed inside the communication system but rather left up to the applications themselves, strictly *communication-specific* functions can be arbitrarily complex. For instance, the end-to-end argument does not prohibit implementing complex routing algorithms within the network.

The underlying reasoning is applicable to not only computer networks but also to systems design in general; for example, the design choices upon which the RISC architecture was built are very similar. When interpreting the last part of the argument ('Sometimes an incomplete version...'), the authors of (Sterbenz et al. 2001) concluded that functions should not be *redundantly* located in the network, but rather replicated where necessary only to improve performance. The end-to-end argument has several facets, and, according to (Saltzer et al. 1998), two complementary goals:

- Higher-level layers, more specific to an application, are free to (and thus expected to) organize lower-level network resources to achieve application-specific design goals efficiently (*application autonomy*).

- Lower-level layers, which support many independent applications, should provide only resources of broad utility across applications, while providing to applications usable means for effective sharing of resources and resolution of resource conflicts (*network transparency*).

If we put this reasoning in the context of some inner network function design (e.g. AQM), we are still left with a plethora of design choices – all we need to do is ensure that our mechanism remains broadly applicable and does not fulfil its purpose for a certain application (or class of applications) only.

2.11.2 Other scalability hazards

Scalability is quite a broad issue; sadly, the end-to-end argument is not all there is to it. Here are some additional scalability hazards:

- *Per-flow state*: This is a major hurdle for making any mechanism scale. If a router needs to identify individual end-to-end flows for any reason (e.g. in order to reserve bandwidth for the flow), it must maintain a table of such flows. In a connectionless network, nodes can normally generate packets at any time and also stop doing so without having to notify routers – thus, the table must be maintained by adding a timer to each entry, refreshing it when a new packet that belongs to this flow is seen, and removing extinct entries when the timer expires. This effort grows linearly with the number of flows in the network, and capacity (processing power or memory) constraints therefore impose an upper limit on the maximum number of flows that can be supported.

As we discussed earlier, traffic phase effects occur because control in routers is executed on a timescale that is not directly related to the RTT of the end-to-end flows in the system. In order to eliminate this problem, a router would need to detect and appropriately treat these individual flows, which requires per-flow state. This explains why it would be hard to come up with a solution for AQM that is significantly better than the currently deployed idea of introducing randomness.

- *Too much traffic*: Even if we assume that the effort of forwarding a packet is fixed and does not depend on the number of flows in the network, router capacity constraints still impose an upper limit on the amount of traffic that can be supported. Hence, traffic that grows significantly with the number of flows can become a scalability hazard. As an example, consider a *peer-to-peer* application (like the file-sharing tools we have all heard of) that requires each involved node to send a lengthy message to all other nodes every second. With two nodes, we have two packets per second. With three nodes, we have six packets per second, and with four nodes, we already have twelve packets per second! Generally, the number of packets per second in the system will be $n(n-1)$, where n is the number of nodes; in the common notation of algorithm runtime complexity, the total traffic therefore scales like $O(n^2)$, which is clearly less scalable than, say, $O(n)$.

It is even possible to eliminate the direct relationship between the amount of network traffic and the number of nodes. This was done, for example, in the case of *RTCP*, which adds signalling of control information to the functionality of the *Real-time Transport Protocol (RTP)* by recommending an upper limit for this type of traffic as a fraction of RTP traffic in the specification (Schulzrinne et al. 2003). The reasoning behind this is that, if RTCP traffic from every node is, say, not more than 5% of RTP traffic, then the whole RTCP traffic in the network will never exceed 5% of the RTP traffic.

- *Poorly distributed traffic*: No matter how the amount of traffic scales with the number of flows, if all the traffic is directed towards a single node, it will be overloaded beyond a certain number of packets (or bytes) per second. Therefore, the traditional *client–server* communication model – a dedicated *server* entity that provides some service to a multitude of *client* entities – is limited in scalability; it has been found that peer-to-peer networks, where each involved entity has an equally important role, and traffic is hence accordingly distributed among several nodes and not directed to a single one, are more scalable. Moreover, having a central point of failure is generally not a good idea.

The potential problem of the client–server model in the last point is frequently exploited in the Internet by the so-called *Distributed Denial-of-Service* (DDoS) attacks. Here, the idea is to flood a server with an enormous amount of traffic from a large number of sources at the same time. The infamous 'TCP SYN' attack takes this idea a step further by additionally exploiting the first scalability hazard in the list: TCP SYN packets, which request a new TCP connection to be opened and therefore lead to per-flow state at the server, are used to flood the node.

To make things worse, the immense scale of the network and the massive amount of traffic in its inner parts renders CPU power costly in Internet backbone routers. This is another fundamental scaling issue that probably has its roots in the topology of the network and its fundamental routing concepts; in any case, you should remember that for a mechanism to be successfully deployed in Internet routers it must not require a lot of processing power.

2.12 Explicit feedback

So far, we have discussed end-system behaviour with simple binary feedback that says 'yes, there was congestion' or 'no, there was no congestion'. We have seen that such feedback can be implicit, that is, based on end-to-end behaviour analysis without any explicit help from within the network. An example of such binary implicit feedback is packet loss: if a packet is only (or mostly) dropped when a queue overflows, its loss indicates congestion. While, as explained in Section 2.4, this is by no means all that can be done with end-to-end measurements, we have seen that it already leads to some interesting issues regarding end-system behaviour. Moreover, there are dynamic effects in the network (phase effects) that should be countered; one solution is AQM, which is difficult to design properly because of all the scalability requirements described in the previous section. All in all, we started out small and simple and now face quite a complex system, which also has a number of potential disadvantages:

- Solely relying on packet drops means that sources always need to increase their rates until queues are full, that is, we have a form of congestion control that first *causes* congestion and then reacts.

- Relying on implicit feedback means making assumptions about the network – interpreting packet loss as a sign of congestion only works well if this is truly the main reason for packets to be dropped.

- We have seen that our system is stable in the case of a fluid-flow model (1 packet = 1 bit) and perfectly equal RTTs. While we have discussed some stabilizing factors that may positively affect the case of heterogeneous RTTs, we did not prove it (and it turns out that a mathematical analysis with heterogeneous RTTs is in fact quite sophisticated (Massoulie 2002)).

Along the way, we gradually extended our model on the basis of logical reasoning and deliberately chose a certain path only twice, and both of these decisions were made in Section 2.4:

1. At the beginning of this section, we postponed considerations regarding aid from within the network and assumed that control is executed at the sender based upon implicit feedback from the receiver.

2. At the end of the section, we decided to refrain from taking more sophisticated implicit feedback into account.

We already left our path a bit when we examined AQM schemes, but even with these mechanisms, help from within the network is not explicit: end systems are not informed about things that happen along the path between them.

2.12.1 Explicit congestion notification

What if they were informed? By dropping packets, AQM already realizes some sort of communication: sources are assumed to increase their rate (and thereby cause the bottleneck queue to grow) until a packet is dropped and then reduce it. Thus, if a packet is dropped earlier, sources will reduce their rates earlier and the queue will grow less significantly. This kind of communication is like standing on someone's feet in order to convey the information 'I do not like you'. Since we do not want to 'hurt the feet' of our sources, an explicit method of communication (*saying* 'I do not like you', or its network equivalent, 'there is congestion') might be a better choice – and indeed, this method works better and does not seem to have any disadvantages whatsoever.

The scheme we are talking about is called *Explicit Congestion Notification (ECN)*, and it is typically realized via a single bit with the semantic '1 = congestion' in packet headers. ECN can be found in a large variety of network technologies, including Frame Relay, ATM and, more recently, IP (Ramakrishnan et al. 2001). Usually, the rule is that routers should set this bit instead of dropping packets when they would normally do so (i.e. when an AQM mechanism would decide to drop the packet), and end systems that see a set ECN bit should update their rate as if the packet had been dropped. The obvious advantage of this method is that it causes less loss – still, a single ECN bit does not suffice for each and every case, as queues can still overflow in the presence of traffic bursts, and this must be reacted upon.

Since the additional effort for routers is almost negligible and does not cause any of the scalability problems that we discussed earlier but nevertheless reduces packet loss, using ECN simply renders AQM more efficient. Moreover, the fact that ECN turns implicit feedback into explicit feedback makes quite a difference: without ECN, end systems are left to guess whether a packet was lost because of an active queue management decision, as a result of a queue that overflowed or because of other unwanted effects (e.g. signal noise that led to a checksum failure and caused the packet to be dropped). With ECN, however, there is a signal available that conveys a clear message and cannot be misinterpreted; moreover, because of its explicit nature, it can be seen by any node that is traversed by packets that carry the signal.

2.12.2 Precise feedback

ECN, which seems to have no real disadvantages but makes things better in a very straightforward manner, has fired the imagination of Internet researchers in recent years. The underlying reasoning of some of their efforts is straightforward: if adding explicit feedback in the form of a single bit does not hurt but makes things better, what would be the result of using *two* bits that would indicate different levels of congestion? What if we even had *three* bits? Also, sources should react to congestion as quickly as possible. Why cannot routers directly inform sources that they should reduce their rates by a certain value in the presence of congestion? Some of the answers to these questions are in fact quite old; the

Table 2.1   Possible combinations for using explicit feedback

Variant no.	Router generates	Router updates	End-system generates
1		×	×
2	×		
3	×		×
4	×	×	
5	×	×	×

questions were asked before. In what follows, we will take a critical look at such explicit feedback methods. As a start, consider Table 2.1, which shows the five basic possibilities to incorporate explicit feedback *from within* the network (note that pure end-to-end feedback from destination to source was not included even though the message itself is explicit). While the table lists 'routers', which is a common name for inner network nodes in the Internet, it applies to other networks too – these nodes could just as well be ATM switches, for example.

The lines in the table are quite generic: an end system can be the source, the destination or both, and routers can generate packets that are sent towards any participating node including other routers. In total, this leaves us with a vast number of possibilities; let us now look at some common examples.

Choke packets

This method (second variant in the table) requires a router to send a message 'please reduce your rate' (perhaps containing further details, such as 'by 50%' or 'to at most 17.3 kbps') to sources directly as soon as it notices congestion (Figure 2.11). Variants of this mechanism, where the notification really says nothing more than 'congestion occurred – reduce your rate', are sometimes called *Backwards ECN* (BECN); an example of this can be found in the Frame Relay standard. Choke packets bring at least two benefits:

1. It is the fastest method to notify sources that they should reduce their rate. This is advantageous because the state of congestion should be changed as quickly as possible.

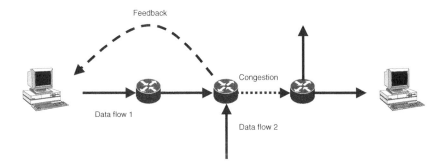

Figure 2.11   Choke packets

2. The entity that knows best about congestion – the bottleneck router – generates congestion feedback, which means that the information is precise.

From a system design perspective, choke packets are, however, somewhat critical. While the disadvantages may be a little less obvious at first sight, they are still significant: first of all, generating a packet is a significant effort for a router. Memory must be allocated, headers must be initialized appropriately and so on – all this happens while the router is overloaded. As explained earlier, CPU power is costly in core Internet routers; mandating that these devices generate packets may not be a good idea, especially when they are congested. Second, this method takes the receiver out of the control loop although the control interval should be dictated by this feedback delay (remember our considerations on the importance of RTT estimation in Section 2.8). It is also a bit critical to send additional packets into the network in a state of congestion, even though these packets go in the other direction. End-to-end ACKs, on the other hand, may be inevitable anyway if a connection-oriented service is desired.

Interestingly, choke packets – which appear to limit the scalability of a network – were the first congestion control suggestion for the Internet (Nagle 1984). The message that was used bears the name *Source Quench* and is conveyed with *ICMP* ('Internet Control Message Protocol') packets. All it says is 'congestion – reduce'. Generating source quench messages is not recommended behaviour for routers anymore (Baker 1995), but it has been brought on the table (only to be pushed off it rather quickly) several times in different disguises since ECN was standardized for the Internet.

Choke packets raise a number of additional questions: how is the congested router supposed to deal with fairness and phase effects, that is, which sources should be notified if packets from thousands of different senders are in the queue? Should it be combined with end-to-end feedback (variants no. 3 and 5 in Table 2.1)? The usefulness of choke packets is perhaps best summarized by this quote from an email by Sally Floyd, which is available from her ECN web page:

> I would agree that there are special cases, mainly high-bandwidth flows over very-long-delay satellite links, where there might be significant benefit to shortening the feedback loop (as would be done with Source Quench). This strikes me as a special, non-typical case, that might require special mechanisms.

Explicit rate feedback

This congestion control method – a realization of variant 1 in Table 2.1 – has the source node generate special packets, or dedicated header fields in regular data packets, which are to be updated by the routers along the path (Figure 2.12). The most prominent example of explicit rate feedback is the ATM *Available Bit Rate (ABR)* service – here, the source generates so-called *Resource Management (RM)* cells, which contain an 'Explicit Rate' field. Each router (these devices are actually called *switches* in ATM terminology, but we stick with the term 'router' for the sake of simplicity) calculates the maximum allowed rate for the sender and reduces the field if its calculated rate is smaller than the value that is already in the field; eventually, the RM cell carries the smallest allowed rate of the path.

The tricky part of this scheme is its feedback calculation in routers. In ATM ABR, for instance, routers are supposed to divide the bandwidth fairly among all flows in the network – in a simplistic implementation, this means detecting, monitoring and counting

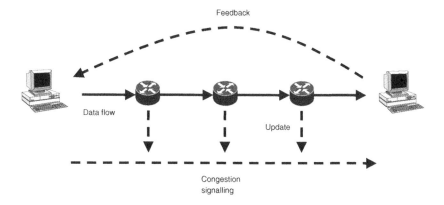

Figure 2.12 Explicit rate feedback

them. This requires per-flow state and is therefore a scalability hazard. Over the years, a wealth of rate-calculation mechanisms for ATM ABR were proposed, all within the rules laid out in the service specification. Some of them are quite sophisticated, and each has its advantages and disadvantages. In general, ATM ABR was abandoned by the Internet community because it did not appear to scale well and required too much effort for routers. This, however, does not necessarily mean that any similar scheme is completely hopeless.

Here is another way of looking at explicit rate feedback: a single ECN bit works fine in the Internet. It is hard to see why two bits, which would inform the source about four different degrees of congestion, would scale much worse. Similarly, it is not so obvious that three or more bits would be a problem. Extend that to a whole 16- or 32-bit field, and you end up with explicit rate feedback, where the signal is a fine-grain congestion notification instead of an allowed rate; this is roughly the opposite information. In any case, the result of this process would look very similar to ATM ABR, but probably show a different rate-calculation method – so, one may wonder whether explicit rate feedback really cannot scale, or if the problem just arose from the rate-calculation mechanisms that were defined then.

As a matter of fact, some very recent suggestions for Internet congestion control resemble explicit rate feedback; we will take a closer look at them in Chapter 4. In addition, Section 3.8 will shed some more light on ATM ABR.

Hop-by-hop congestion control

In such a scheme (sometimes also called *backpressure*, each router (hop) along an end-to-end path sends feedback to the directly preceding router, which executes control based on this feedback (Figure 2.13). Such schemes were studied in great detail in the past (Gerla

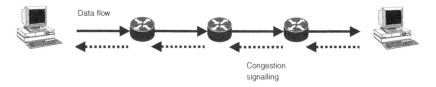

Figure 2.13 Hop-by-hop congestion control

and Kleinrock 1980) but received minor attention in the literature in recent years, which is perhaps owing to the significant amount of work that they impose on routers. These days, congestion control is mainly an Internet issue, and even explicit rate feedback schemes appear to cause too much router overhead for them to become deployed. It is hard to imagine that a hop-by-hop scheme would require as much, or less, per-hop work than an explicit rate feedback scheme.

In general, the goal of hop-by-hop schemes was to attain reliability *inside* the network, for example, by buffering packets at each hop until they can be sent on (because there is no more congestion) and only transmitting as much as the next hop allows; this is a dream that Internet designers have abandoned a long time ago. In addition to the problems that seem more obvious to us nowadays (scalability, buffer requirements, etc), researchers had to deal with the following peculiar issue: the notion of a reliable end-to-end data stream can lead to *deadlocks*, where router A waits for the buffer of router B to drain, router B waits for the buffer of router C to drain and router C waits until the buffer in A is empty.

Despite all its problems, hop-by-hop congestion control is an interesting concept because it has the potential to give all involved entities the power to execute very immediate control. Some researchers still look at it, albeit in special scenarios only – for example, wireless multi-hop networks (Yung and Shakkottai 2004).

Concluding remarks

We have discussed the most common, but by no means all, possibilities to utilize explicit feedback from routers. A realization of variant 5 in Table 2.1 could, for example, combine explicit rate feedback with choke packets (this is part of the ATM ABR design). In reality, some combinations might in fact be inevitable: for instance, in order to gradually deploy an explicit rate feedback mechanism in the Internet (where pure end-to-end feedback currently dominates), relying on nothing but explicit help from routers may lead to a mechanism that never really works because there are always some routers that do not implement the new scheme in the system. If, in such a case, the bottleneck router does not understand explicit rate feedback, the whole mechanism is pointless.

While the Internet community does not really embrace congestion control methods that involve explicit feedback because of the critical extra work for routers and some other reasons such as poor interactions with *IPSec* tunnels,[5] the success of ECN and the increasing number of problems experienced with existing mechanisms as the nature of network connections changes (e.g. capacities have grown significantly, and more and more users now have a wireless Internet connection) have sparked interest. In what follows, we will first describe the impact that heterogeneous environments can have in the absence of explicit help from within the network, and then discuss the fundamental design problem that causes these issues.

2.13   Special environments

So far, our observations involved end systems, packets, links that have certain capacity and delay properties, and routers with queues. The Internet was also mentioned several times, as it is the most-important operational network that exhibits both the necessity and the

---

[5]IPSec is a protocol suite that provides end-to-end security at the network level – see RFC 2401 (Kent and Atkinson 1998) for further details.

solutions for congestion control. While it is well known that Internet access now spans a broad range of devices and environments (from the cell phone in my pocket to the satellite way above me – in fact, there is now even research on the 'Interplanetary Internet'), we have totally neglected this aspect of network communications up to now. However, it does play a significant role for congestion control.

Briefly put, the influence of heterogeneous environments can falsify the interpretation of network measurements. As already mentioned in Section 2.4, only three things can happen to a single packet as it traverses the network: (i) it can be delayed, (ii) it can be dropped, and (iii) it can be changed. At least two of these effects lend themselves to interpretations by an end-to-end receiver for the sake of providing the sender with feedback about the state of congestion in the network: delay, which can be a result of growing queues, and packet drops, which can be a result of an overflowing queue. We have also seen that, whether increasing delay is interpreted as a sign of congestion or not, measuring the RTT is useful for several things. Here is a brief (non comprehensive) list of different environments with their properties that have an influence on end-to-end delay and loss:

Wireless links: In wireless usage scenarios, bit errors are frequent; they can, for instance, occur when a user passes by a wall or enters an elevator. At the wireless receiver, receiving noisy data means that a checksum will fail. Depending on the specific technology that is used, this can mean two things:

1. The packet is dropped (because its contents are now useless).

2. The packet is delayed, because the link layer retransmits it if an error occurs. Even then, if multiple retransmissions fail, the packet can eventually be dropped.

Typically, bit errors are not statistically independent, that is, they are clustered in time (Karn et al. 2004).

Satellite links: Satellites combine a number of potentially negative factors:

- Connections are noisy, which means that they share the negative properties of wireless links described above. This problem may be compensated for with *Forward Error Correction* (FEC), but this can add delay.

- They show a long delay, which is always a problem because of its impact on the precision of RTT measurements: the larger the delay, the larger its fluctuations. In the case of *GEOs* – satellites in geostationary orbit – the delay is between 239.6 and 279.0 ms (Allman et al. 1999a). While 'low earth orbit' satellites (*LEOs*) seem to alleviate this problem via their significantly reduced one-way propagation delay of about 20 to 25 ms, it may be necessary to route data *between* them ('LEO handoff'), which can cause packet loss and propagation delay fluctuations (Hassan and Jain 2004).

- Satellites are typical examples of so-called long fat pipes – links with a large bandwidth $\times$ delay product. As we have discussed earlier, this value dictates the amount of data a protocol should have 'in flight' (transmitted, but not yet acknowledged) in order to fully utilize a link (Allman et al. 1999a). Such links cause problems for mechanisms like AIMD: additively increasing the rate with a constant factor until the link is saturated can take a very long time (perhaps too long, as there should be no loss during the increase phase until the limit is reached) while the bandwidth reduction due to multiplicative decrease is quite drastic.

- Another common problem is that some Internet providers offer a satellite down-link, but the end-user's outgoing traffic is still sent over a slow terrestrial link (or a satellite uplink with reduced capacity). With window-based congestion control schemes, this highly asymmetric kind of usage can cause a problem called *ACK starvation* or *ACK congestion*, in which the sender cannot fill the satellite channel in a timely fashion because of slow acknowledgements on the return path (Metz 1999). As we have seen in Section 2.7, enqueuing ACKs because of congestion can also lead to traffic bursts if the control is window based.

Mobility: As users move from one access point (base station, cell... depending on the technology in use) to another while desiring permanent connectivity, two noteworthy problems occur:

1. Normally, any kind of link layer technology requires a certain time period for handoff (during which no packets can be transmitted) before normal operation can continue.

2. If the moving device is using an Internet connection, which should be maintained, it should keep the same IP address. Therefore, mechanisms for *Mobile IP* come into play, which may require incoming packets to be forwarded via a 'Home Agent' to the new location (Perkins 2002). This means that packets that are directed to the mobile host experience increased delay, which has an adverse effect on RTT estimation.

This list contains only a small subset of network environments – there are a large number of other technologies that roughly have a similar influence on delay and packet loss. ADSL connections, for example, are highly asymmetric and therefore exhibit the problems that were explained above for direct end user satellite connections. DWDM-based networks using optical burst or packet switching may add delay overhead or even drop packets, depending on the path setup and network load conditions (Hassan and Jain 2004). Some of the effects of mobility and wireless connections may be amplified in mobile ad hoc networks, where connections are in constant flux and customized routing schemes may cause delay overhead. Even much more 'traditional' network environments show properties that might have an adverse effect on congestion control – for example, link layer MAC functions such as Ethernet CSMA/CD can add delay, and so can path changes in the Internet.

2.14   Congestion control and OSI layers

When I ask my colleagues where they would place congestion control in the OSI model, half of them say that it has to be layer 3, whereas the other half votes for layer 4. As a matter of fact, a Google search on 'OSI' and 'congestion control' yields quite similar results – it gives me documents such as lecture slides, networking introduction pages and so on, half of which place the function in layer 4 while the other half places it in layer 3. Why this confusion?

The standard (ISO 1994) explicitly lists 'flow control' as one of the functions that is to be provided by the network layer. Since this layer is concerned with intermediate systems, the term 'flow control' cannot mean slowing down the sender (at the endpoint) in order to protect the receiver (at the endpoint) from overload in this context; rather, it means slowing down intermediate senders in order to protect intermediate receivers from overload (as explained on Page 14, the terms 'flow control' and 'congestion control' are sometimes used synonymously). Controlling the rate of a data flow within the network for the sake of the network itself is clearly what we nowadays call 'congestion control'.

Interestingly, no such function is listed for the transport layer in (ISO 1994) – but almost any introductory networking book will (correctly) tell you that TCP is a transport layer protocol. Also, TCP is the main entity that realizes congestion control in the Internet – complementary AQM mechanisms are helpful but play a less crucial role in practice. Indeed, embedding congestion control in TCP was a violation of the ISO/OSI model. This is not too unusual, as Internet protocols, in general, do not strictly follow the OSI rules – as an example, there is nothing wrong with skipping layers in TCP/IP. One reason for this is that the first Internet protocol standards are simply older than ISO/OSI. The important question on the table is, Was it good design to place congestion control in the transport rather than in the network layer?

In order to answer this, we need to look at the reason for making 'flow control' a layer 3 function in the OSI standard: congestion occurs *inside* the network, and (ISO 1994) explicitly says that the network layer is supposed to hide details concerning the inner network from the transport layer. Therefore, adding functionality in the transport layer that deduces implicit feedback from measurements based on assumptions about lower layers (e.g. packets will mainly be dropped as a result of congestion) means to work against the underlying reasoning of the OSI model.

The unifying element of the Internet is said to be the IP datagram; it is a simple intermediate block that can act as a binding network layer element between an enormous number of different technologies on top and underneath. The catchphrase that says it all is: 'IP over everything, everything over IP'. Now, as soon as some technology on top of IP makes implicit assumptions about lower layers, this narrows the field of usability somewhat – which is why we are facing well-known problems with TCP over heterogeneous network infrastructures. Researchers have come up with a plethora of individual TCP tweaks that enhance its behaviour in different environments, but there is one major problem here: owing to the wide acceptance of the whole TCP/IP suite, the binding element is no longer just IP but it is, in fact, TCP/IP – in other words, you will need to be compatible with legacy TCP implementations, or you cannot speak with thy neighbour. Today, 'IP over everything, everything over TCP' is more like it.

2.14.1 Circuits as a hindrance

Van Jacobson made a strong point against building circuits into the Internet, during his keynote speech[6] at the ACM SIGCOMM 2001 conference in San Diego, California. He explained how we all learned, back in our schooldays, that circuits are a simple and

---

[6]At the time of writing, the slides were available from http://www.acm.org/sigcomm

fundamental concept (because this is how the telephone works), whereas in fact, the telephone system is more complex and (depending on its size) less reliable than an IP-based network. Instead of realizing circuits on top of a packet-based best effort network, we should perhaps strive towards a network that resembles the power grid. When we switch on the light, we do not care where the power comes from; neither does a user care about the origin of the data that are visualized in the browser upon entering, say, `http://www.moonhoax.com`. However, this request is normally associated with an IP address and a circuit is set up.

It is a myth that the Internet routes around congestion; it does not. Packets do not individually find the best path to the destination on the basis of traffic dynamics – in general, a path is decided for and kept for the duration of a connection unless a link goes down. Why is this so? As we will see in Chapter 5, ISPs go to great lengths to properly distribute traffic across their networks and thereby make efficient use of their capacities; these mechanisms, however, are circuit oriented and hardly distribute packets individually – rather, decisions are made on a broader scale (that is, on a user aggregate, user or at least connection basis).

Appropriately bypassing congestion on a per-packet basis would mean that packets belonging to the same TCP connection could alternate between, say, three different paths, each yielding different delay and loss behaviour. Then, making assumptions about the network would be pointless (e.g. while reacting to packet loss might seem necessary, the path could just have changed, and all of a sudden, there could be a perfectly congestion free situation in the network). Also, RTT estimation would suffer, as it would no longer estimate the RTT of a given connection but rather follow an average that represents a set of paths. All in all, the nature of TCP – the fact that it makes implicit assumptions about lower layers – mandates that a path remain intact for a while (ideally the duration of the connection).

Theoretically, there would be two possibilities for solving this problem: (i) realizing congestion control in layer 3 and nowhere else, or (ii) exclusively relying on explicit feedback from within the network. The first approach would lead to hop-by-hop congestion control strategies, which, as we have already discussed, are problematic for various reasons. The latter could resemble explicit rate feedback or use choke packets, but again, there are some well-known issues with each of these methods. The Internet approach of relying on implicit assumptions about the inner network, however, has proved immensely scalable and reached worldwide success despite its aforementioned issues.

It is easy to criticize a design without providing a better solution; the intention of this discussion was not to destructively downplay the value of congestion control as it is implemented in the Internet today, but to provide you with some food for thought. Chances are that you are a Ph.D. student, in which case you are bound to be on the lookout for unresolved problems – well, here is a significant one.

2.15 Multicast congestion control

In addition to the variety of environments that make a difference for congestion control mechanisms, there are also network operation modes that go beyond the relatively simple *unicast* scenario, where a single sender communicates with a single receiver. Figure 2.14 illustrates some of them, namely, *broadcast*, *overlay multicast* and network layer *multicast*; here, 'S' denotes a sender and 'R' denotes receivers. The idea behind all of these

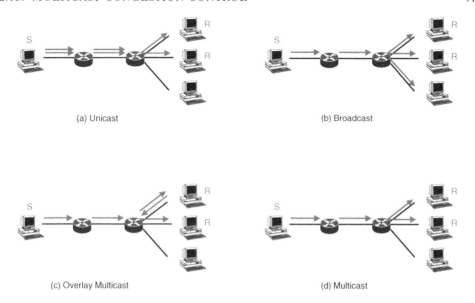

Figure 2.14 Unicast, broadcast, overlay multicast and multicast

communication modes is that there are multiple receivers for a stream that originates from a single sender – for example, a live radio transmission. In general, such scenarios are mostly relevant for real-time multimedia communication.

The reason why multicast differs from – and is more efficient than – unicast can be seen in Figure 2.14 (a): in this diagram, the stream is transmitted twice across the first two links, thereby wasting bandwidth and increasing the chance for congestion. Multicast (Figure 2.14 (d)) solves this by having the second router distribute the stream towards the receivers that participate in the session. In this way, multicast constructs a *tree* instead of a single end-to-end path between the sender and receivers.

The other two communication modes are shown for the sake of completeness: broadcast is what actually happens with radio transmission – whether you are interested or not, your radio receives transmissions from all radio stations in range. It is up to you to apply a filter based on your liking by tuning the knob (selecting a frequency). The figure shows that this is inefficient because the bandwidth from the second router to the lower end system that does not really want to participate is wasted.

Overlay multicast (Figure 2.14 (c)) is what happens quite frequently as an interim solution while IP multicast still awaits global deployment: some end systems act like inter-mediate systems and take over the job that is supposed to be done by routers. The diagram shows that this is not as efficient as multicasting at the network layer: this time, bandwidth from the upper receiver back to the second router is wasted. Clearly, multicast is the winner here; it is easy to imagine that the negative effects of the other transmission modes would be much more pronounced in larger scenarios (consider, for example, a sender and 100 receivers in unicast mode, or a large tree that is flooded in broadcast mode). The bad news is that congestion is quite difficult to control in this transmission mode.

2.15.1 Problems

Let us take a look at two of the important issues with multicast congestion control that were identified by the authors of (Yang and Lam 2000b):

Feedback implosion: If a large number of receivers independently send feedback to a single receiver, the cumulative amount of such signalling traffic increases as it moves upwards in the multicast tree. In other words, links that are close to the sender can become congested with a massive amount of feedback. This problem does not only occur with congestion control specific feedback: things are no better if receivers send ACKs in order to realize reliable communication.

This problem can be solved by suppressing some of the feedback. For example, some receivers that are chosen as representatives could be the only ones entitled to send feedback; this method brings about the problem of finding the right criteria for selecting representatives. Instead of trying to pick the most-important receivers, one could also limit the amount of signalling traffic by other means – for example, by controlling it with random timers. Another common class of solutions for the feedback implosion problem relies on *aggregation*. Here, receivers do not send their feedback directly to the sender but send it to the first upstream router – an inner node in the multicast tree. This router uses the information from multiple feedback messages to calculate the contents for a single collective feedback message. Each router does so, thereby reducing the number of signalling messages as feedback moves up in the tree.

Feedback filtering and heterogeneous receivers: No matter how (or if) the feedback implosion problem is solved, multicasting a stream implies that there will be multiple independent receivers that potentially experience a different quality. This depends not only on the receiving device but also on the specific branch of the tree that was traversed – for example, congestion might occur close to the source, right in the middle of the tree or close to the receiver. Link bandwidths can vary. Depending on the performance they see, the multicast receivers will provide different feedback. A single packet may have been lost along the way to two receivers but it may have successfully reached three others. What should be done? Is it worth retransmitting the packet?

Clearly, a filter function of some sort needs to be applied. How this is done relates to the solution of the feedback suppression problem: if feedback is aggregated, the intermediate systems that carry out the aggregation must somehow calculate reasonable collective feedback from the individual messages they receive. Thus, in this case, the filter function is distributed among these nodes. If feedback suppression is solved by choosing representatives, this automatically means that feedback from these receivers (and no others) will be taken into account. The problem of choosing the right representative remains.

There are still some possibilities to cope with the variety of feedback even if we neglect feedback implosion: for instance, the sender could use a timer that is based on an average RTT in the tree and only react to feedback once per timer interval.

In order to avoid phase effects and amply satisfy all receivers, this interval could depend upon a random function. The choice also depends on the goals: is it more important to provide good quality on average, or is it more important that no single receiver experiences intolerable quality? In the latter case, it might seem reasonable to dynamically choose the lossiest receiver as the representative.

2.15.2 Sender- and receiver-based schemes

The multicast congestion control schemes we have considered so far are called *sender-based* or *single-rate* schemes because the sender always decides to use a certain single rate for all receivers of the stream at the sender. *Layered* (*receiver-based, multi-rate*) schemes follow a fundamentally different approach: here, the stream is hierarchically encoded, and it is up to the receivers to make a choice about the number of layers that they can cope with. This obviously imposes some requirements on the data that are transmitted – for example, it would not make much sense for reliable file transfer. Multimedia data, however, may sometimes be ordered according to their importance, thereby rendering the use of layers feasible.

One such example is *progressive encoding* of JPEG images: if you remember the early days of Internet surfing, you might recall that sometimes an image was shown in the browser with a poor quality at first, only to be gradually refined afterwards. The idea of this is to give the user a first glance of what an image is all about, which might lead to a quick choice of interrupting the download instead of having to wait in vain. Growing Internet access speeds and, perhaps also, web design standards have apparently rendered this technically reasonable but visually not too appealing function unfashionable. There is also the disadvantage that progressive JPEG encoding comes at the cost of increasing the total image size a bit. In a multicast setting, such a function is still of interest: a participant could choose to receive only the data necessary for minimal image quality and refrain from downloading the refinement part. In reality, the data format of concern is normally not JPEG but often an audio or video stream. The latter, in particular, received a lot of attention in the literature (Matrawy and Lambadaris 2003).

A receiver informs the sender (or upstream routers) which layers it wants to receive via some form of signalling. As an example, the sender could transmit certain layers to certain *multicast groups* only – collections of receivers that share common properties such as interest in a particular layer – and a receiver could inform the sender that it wants to join or leave a group. The prioritization introduced by separating data into layers can be used for diverse things in routers; for instance, an AQM scheme could assign a higher dropping priority to packets that belong to a less important layer, or routers could refrain from forwarding packets to receivers that are not interested in them altogether. This, of course, raises scalability concerns; one must find a reasonable trade-off between efficient operation of a multicast congestion control scheme and requiring additional work for routers.

While it is clear from Figure 2.14 that multicast is the most-efficient transmission mode whenever there is one sender and several receivers, there are many more problems with it than we have discussed here. As an example, fairness is quite a significant issue in this context. We will take a closer look at it towards the end of this chapter – but let us consider the role of incentives first.

2.16   Incentive issues

So far, we have assumed that all entities that are involved in a congestion control scheme are willing to cooperate, that is, adhere to the rules prescribed by a scheme. Consider Figure 2.4 on Page 17: what would the trajectories look like if only customer 0 implements the rate update strategy and customer 1 simply keeps sending at the greatest possible rate? As soon as customer 0 increases its rate, congestion would occur, leading customer 0 to reduce the rate again. Eventually, customer 0 would end up with almost no throughput, whereas customer 1, which greedily takes it all, obtains full capacity usage. Thus, if we assume that every customer selfishly strives to maximize its benefit by acting in an unco-operative manner, congestion control as we have discussed cannot be feasible. Moreover, such behaviour is not only unfair but also inefficient – as we have seen in the beginning of this chapter, under special circumstances, total throughput through a network can decrease if users recklessly increase their sending rates.

2.16.1   Tragedy of the commons

In the Internet, network capacity is a common resource that is shared among largely indepen-dent individuals (its users). As stated in a famous science article (Hardin 1968), uncontrolled use of something that everybody can access will only lead to ruin (literally, the article says that 'freedom in a commons brings ruin to all'). This is called the *tragedy of the commons*, and it develops as follows: Consider a grassy pasture, and three herdsmen who share it. Each of them has a couple of animals, and there is no problem – there is enough grass for everybody. Some day, one of the herdsmen may wonder whether it would be a good idea to add another animal to his herd. The logical answer is a definite yes, because the utility of adding an animal is greater than the potential negative impact of overgrazing from the single herdsman's point of view. Adding an animal has a direct positive result, whereas overgrazing affects all the herdsmen and has a relatively minor effect on each of them: the total effect divided by the number of individuals. This conclusion is reached by any herdsman at any time – thus, all herds grow in size until the pasture is depleted.

The article, which is certainly not without controversy, goes on to explain all kinds of commonly known society problems by applying the same logic, ranging from the nuclear arms race to pollution and especially overpopulation. In any case, it appears reasonable to apply this logic to computer networks; this was done in (Floyd and Fall 1999), which illus-trates the potential for disastrous network-wide effects that unresponsive (selfish) sources can have in the Internet, where most of the traffic consists of congestion controlled flows (Fomenkov et al. 2004). One logical conclusion from this is that we would need to *reg-ulate* as suggested in (Hardin 1968), that is, install mechanisms in routers that prevent uncooperative behaviour, much like traffic lights, which prevent car crashes via regulation.

2.16.2   Game theory

This scenario – users who are assumed to be uncooperative, and regulation inside the net-work – was analysed in (Shenker 1994); the most-important contribution of this work is perhaps not the actual result (the examined scenario is very simplistic and the assumed Poisson traffic distribution of sources differs from what is found in the Internet) but rather the extensive use of game theory as a means to analyse a computer network. Game-theoretic models for networks have since become quite common because they are designed to answer a question that is important in this context: how to optimize a system that is comprised of uncooperative and selfish users. The approach in (Shenker 1994) is to consider *Nash equilibria* – sets of user strategies where no user has an incentive to change her strategy – and examine whether they are efficient, fair, unique and easily reachable. Interestingly, although it is the underlying assumption of a Nash equilibrium that users always strive to maximize their own utility, a Nash equilibrium does not necessarily have to be efficient.

The tragedy of the commons is a good example of an inefficient Nash equilibrium: if, in the above example, all herdsmen keep buying animals until they are out of money, they reach a point where none of them would have an incentive to change the situation ('if I sell a cow, my neighbour will immediately fill the space where it stood') but the total utility is not very high. The set of herdsmen's strategies should instead be *Pareto optimal*, that is, maximize their total utility.[7] In other words, there would be just as many animals as the pasture can nourish. If all these animals belong to a single farmer, this condition is fulfilled – hence, the goal must be a *fair* Pareto optimal Nash equilibrium, where the ideal number of animals is equally divided among the herdsmen and none of them would want to change the situation. As a matter of fact, they *would* want to change it because the original assumption of our example was that the benefit from buying an animal would outweigh the individual negative effect of overgrazing; thus, in order to turn this situation into a Nash equilibrium, a herdsman would have to be punished by some external means if he was to increase the size of his herd beyond a specified maximum. This necessity of regulation is the conclusion that was reached by in (Hardin 1968), and it is also one of the findings in (Shenker 1994): simply servicing all flows with a FIFO queue does not suffice to ensure that all Nash equilibria are Pareto optimal and fair.

2.16.3 Congestion pricing

Punishment can take different forms. One very direct method is to tune router behaviour such that users can attain maximum utility (performance) only by striving towards this ideal situation; for the simple example scenario in (Shenker 1994), this means changing the queuing discipline. An entirely different possibility (which is quite similar to the notion of punishment in our herdsmen–pasture example) is *congestion pricing*. Here, the idea is to alleviate congestion by demanding money from users who contribute to it. This concept is essentially the same as congestion charging in London, where drivers pay a fee for entering the inner city during times of expected traffic jams.[8] Economists call costs of a good that do not accrue to the consumer of the good as *externalities* – social issues such as pollution or traffic congestion are examples of negative externalities. In economic terms, congestion charging is a way to *internalize the externalities* (Henderson et al. 2001).

Note that there is a significant difference between the inner city of London and the Internet (well, there are several, but this one is of current concern to us): other than an ISP, London is not a player in a free market (or, if you really want to see it that way, it is a player that imposes very high opportunity costs on customers who want to switch to another

---

[7]In a Pareto optimal set of strategies, it is impossible to increase a player's utility by changing the strategy without decreasing the utility of another player.

[8]7 a.m. to 6.30 p.m. according to http://www.cclondon.com at the time of writing.

one – they have to leave the city). In other words, in London, you just have to live with congestion charging, like it or not, whereas in the Internet, you can always choose another ISP. This may be the main reason why network congestion–based charging did not reach wide acceptance – but there may also be others. For one, Internet congestion is hard to predict; if a user does not know in advance that the price will be raised, this will normally lead to utter frustration, or even anger. Of course, this depends on the granularity and timescale of congestion that is considered: in London, simple peak hours were defined, and the same could be done for the Internet, where it is known that traffic normally increases in the morning before everybody starts to work. Sadly, such a form of network congestion pricing is of minor interest because the loss of granularity comes at the cost of the main advantage: *automatic network stability via market self regulation.*

This idea is as simple as it is intriguing: it a well-known fact that a market can be stabilized by controlling the equilibrium between supply and demand. Hence, with well-managed congestion pricing, there would be no need for stabilizing mechanisms inside the network: users could be left to follow their own interests, no restraining mechanisms would need to be deployed (neither in routers nor in end systems), and all problems could be taken care of by amply charging users who contribute to congestion. Understandably, this concept – and the general unbroken interest in earning money – led to a great number of research efforts, including the European M3I ('Marked Managed Multiservice Internet') project.[9] The extent of all this work is way beyond the scope of this book, including a wealth of consideration that are of an almost purely economic nature; (Courcoubetis and Weber 2003) provides comprehensive coverage of these things. Instead of fully delving into the depth of this field, let us briefly examine a famous example that convincingly illustrates how the idea of congestion pricing manages to build a highly interesting link between economy and network technology: the *smart market.*

This idea, which was introduced in (MacKie-Mason and Varian 1993), works as follows: consider users that participate in an auction for the bandwidth of a single link. In this auction, there is a notion of discrete time slots. Each packet carries a 'bid' in its header (the price the user is willing to pay for transmission of the packet), and the network, which can transmit m out of n packets, chooses the m packets with the highest bids. It is assumed that users would normally set default bids for various applications and only change them under special circumstances (i.e. for very bandwidth- and latency-sensitive or insensitive traffic, as generated by an Internet telephony or an email client, respectively). The price to be charged to the user is the highest bid found in packets that did not make it; this price is called *marginal cost* – the cost of sending one additional packet. The reason for this choice is that the price should equal marginal cost for the market to be in equilibrium. Other than most strictly technical solutions, this scheme has the potential to stabilize the network without requiring cooperative behaviour because each user gains the most benefit by specifying exactly the amount that equals her true utility for bandwidth (Courcoubetis and Weber 2003).

Albeit theoretically appealing, the smart market scheme is generally known to be impractical (Henderson et al. 2001): it is designed under the assumption that packets encounter only a single congested link along the path; moreover, it would require substantial hardware investments to make a router capable of all this. Such a scheme must, for instance, be accompanied by an efficient, scalable and secure signalling protocol. As mentioned above,

---

[9]http://www.m3i.org/

other forms of network congestion pricing may not have reached the degree of acceptance that researchers and perhaps also some ISPs hope for because of the inability to predict congestion. Also, a certain reluctance towards complicated pricing schemes may just be a part of human nature.

The impact of incentives and other facets of human behaviour on networks (and, in particular, the Internet) is still a major research topic, where several questions remain unanswered and problems are yet to be tackled. For instance, (Akella et al. 2002) contains a game-theoretic analysis of congestion control in the Internet of today and shows that it is quite vulnerable to selfish user behaviour. There is therefore a certain danger that things may quickly change for the worse unless an incentive compatible and feasible congestion control framework is put into place soon. On the other hand, even the underlying game-theoretic model itself may need some work – the authors of (Christin et al. 2004) point out that there might be more-suitable notions of equilibrium for this kind of analyses than pure Nash equilibria.

2.17 Fairness

Let us now assume that all users are fully cooperative. Even then, we can find that the question of how much bandwidth to allocate to which user has another facet – *fairness*. How to fairly divide resources is a topic that mathematicians, lawyers and even philosophers like Aristotele have dealt with for a long time. Fairness is easy as long as everybody demands the same resources and asserts a similar claim – as soon as we relax these constraints, things become difficult. The Talmud, the monumental work of Jewry, explains a related law via an example that very roughly goes as follows:

> *If two people hold a cloth, one of them claims that it belongs to him, and the other one claims that half of it belongs to him, then the one who claims the full cloth should receive 3/4 and the other one should receive 1/4 of the cloth.*

This is in conflict with the decision that Aristotele would have made – he would have given 2/3 to the first person and 1/3 to the other (Balinski 2004). Interestingly, both choices appear to be inherently reasonable: in the first case, each person simply shares the claimed part. In the second case, the cloth is divided proportionally, and the person who claimed twice as much receives twice the amount of the other one.

In a network, having users assert different claims would mean that one user is more important than the other, perhaps because she paid more. This kind of per-user prioritization is a bit of a long lost dream in the history of computer networks. It was called *Quality of Service (QoS)*, large amounts of money and effort were put into it and, bluntly put, nothing happened (we will take a much more detailed look at QoS in Chapter 5, but for now, this is all you need to know). Therefore, in practice, difficulties of fairness only arise when users do not share the similar (or the same amount of) resources. In other words, all the methods to define fairness that we will discuss here would equally divide an apple among all users if an apple is all we would care about. Similarly, in Figure 2.4, defining fairness is trivial because there is only one resource.

Before we go into details of more complex scenarios, it is perhaps worth mentioning how equal sharing of a single resource can be quantified. This can be done by means of *Raj*

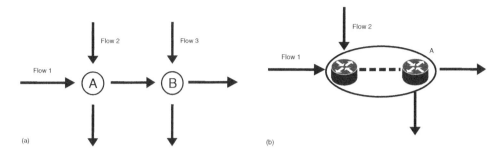

Figure 2.15 Scenario for illustrating fairness (a); zooming in on resource A (b)

Jain's fairness index (Jain et al. 1984): if the system allocates rates to n contending users, such that the ith user receives a rate allocation x_i, the fairness index $f(x)$ is defined as

$$f(x) = \frac{\left(\sum_{i=1}^{n} x_i\right)^2}{n \sum_{i=1}^{n} x_i^2} \tag{2.2}$$

This is a good measure for fairness because $f(x)$ is 1 if all allocations x_i are perfectly equal and immediately becomes less than 1 upon the slightest deviation. It is therefore often used to evaluate the fairness of congestion control mechanisms.

Why are things more complex when users share different resources? For one, the sending rate of each flow in a network is determined by its respective bottleneck link. If the bottleneck is not the same for all flows, simply partitioning all bottleneck links fairly may lead to an inefficient rate allocation. To understand what this means, look at Figure 2.15: in (a), we have three flows that share two resources. Depending on what needs to be shared, a resource can be, for example, a router or an intermediate link (the dashed line in (b)). We will focus on the latter case from now on. Let us assume that resources A and B constrain the bandwidth to 10 and 20 kbps, respectively. Then, fairly sharing resource A would mean that flows 1 and 2 both attain a rate of 5 kbps. Fairly dividing the capacity of resource B among flows 1 and 3 would yield 10 kbps each, but the bandwidth of flow 1 is already constrained to a maximum of 5 kbps by resource A – it cannot use as much of resource B as it is given.

2.17.1 Max–min fairness

It is obvious that simply dividing single resources is not good enough, as it can lead to a rate allocation where several users could increase the rate even further without degrading the throughput of others. In our example, there is unused capacity in B (the 5 kbps that cannot be used by flow 1); there would be no harm in allowing flow 3 to increase its rate until it reaches the capacity limit with a rate of 15 kbps. In a more general form, a method to attain such a rate allocation can be written as follows:

- At the beginning, all rates are 0.

- All rates grow at the same pace, until one or several link capacity limits are hit.
  (Now we have reached a point where each flow is given a fair share of at least one bottleneck. Yet, it is possible to achieve a better utilization.)

- All rates except for the rates of flows that have already hit a limit are increased further.

- This procedure continues until it is not possible to increase.

This is the *progressive filling algorithm*: It can be shown to lead to a rate allocation that is called *max–min fairness*, which is defined as follows (Le Boudec 2001):

> *A feasible allocation of rates $\vec{x}$ is 'max–min fair' if and only if an increase of any rate within the domain of feasible allocations must be at the cost of a decrease of some already smaller or equal rate. Formally, for any other feasible allocation $\vec{y}$, if $y_s > x_s$ then there must exist some s' such that $x_{s'} <= x_s$ and $y_{s'} < x_{s'}$.*

In this context, a *feasible* rate allocation is, informally, a rate allocation that does not exceed the total network capacities and is greater than 0. This definition basically says that a max–min fair rate allocation is Pareto optimal. Another way to define this rate allocation is:

> *A feasible allocation of rates $\vec{x}$ is 'max–min fair' if and only if every flow has a bottleneck link (a link that limits its rate).*

A max–min fair rate allocation is unique; this definition of fairness was very prominent for a long time and was recommended in the ATM *ABT* specification (see Section 3.8). Its name stems from the fact that, in a sense, it favours flows with smaller rates. Even though max–min fairness seems to provide an acceptable rate allocation vector at first sight, it may not lead to perfect network utilization because resources are divided irrespective of the number of resources that a flow consumes. If, for instance, resources A and B in Figure 2.15 would each have the same capacity c and we were to maximize the total throughput of the network, we would assign the rate c to flows 1 and 2 and a rate of zero to flow 3, yielding a total network throughput of $2c$. Yet, in the case of max–min fairness, the total throughput would only be $\frac{3}{2}c$. Does this mean that a totally unfair allocation makes more sense because it uses the available resources more efficiently?

2.17.2 Utility functions

At this point, it may be wise to reconsider the goal of congestion control: *we want to avoid congestion*. The reason for this is, arguably, that *we want to provide users with the best possible service*. In economic terms, user experience can be expressed as the amount a user is willing to pay. Therefore, we should ask: even if the rate vector (*flow*1 = 0, *flow*2 = c, *flow*3 = c) maximizes network throughput, would we be able to earn the greatest amount of money with this kind of rate distribution? According to (Shenker 1995), the answer is 'no'. In the case of traditional Internet usage, for instance, (these are so-called *elastic applications* such as file transfer or web browsing) each flow can be associated with an increasing, strictly concave and continuously differentiable *utility function*. In other words, the utility of additional bandwidth – the willingness to pay for it – decreases as the bandwidth granted to a user increases.

This is a common economic phenomenon: if I enter a supermarket and plan to buy a single box of detergent but there is an offer to obtain a second box for half the price provided that two boxes are bought at the same time, I am usually tempted to accept this offer. The constant presence of such offers indicates the success of these schemes, and it could perhaps even be taken one step further: maybe it is just me, but if a third box would cost half as much as the second one, I am pretty sure that I would even buy three boxes. For me, this is only limited by the total weight that I can carry home. Whether such a strategy works obviously depends on the product: sometimes, the same is done with milk that expires the next day, but since I can make use of at most one litre per day, this is all that I will buy.

Bandwidth for file transfer or web browsing is quite similar to detergent: the price that a user will be willing to pay to obtain a throughput of 15 Mbps instead of 10 Mbps is typically less than the price the user will pay to obtain 10 Mbps instead of 5 Mbps. There is hardly a limit – in general, the more the bandwidth, the better it is. Other applications, such as telephony, are more like milk: a certain amount of bandwidth is necessary for the application to work, but anything beyond a certain limit is useless. Such properties of goods, or applications, can be expressed with their utility functions. Typical examples from (Shenker 1995) are shown in Figure 2.16: in addition to the already discussed *elastic* applications, there are applications with *hard real-time* constraints. These applications simply do not function if the bandwidth is below a certain limit and they work fine as soon as the limit is reached. Anything beyond the limit does not yield any additional benefit.

The other two types shown in the figure are *delay-adaptive* applications (applications with *soft real-time* constraints) and *rate-adaptive* applications. The curve of delay-adaptive applications more closely matches the utility of a regular Internet real-time application, such as a video conferencing or streaming audio tool. These applications are typically designed to work across the Internet, that is, they tolerate a certain number of occasional delay bound violations or even some loss. Still, if the bandwidth is very low, such an application is normally useless. Rate-adaptive applications (often just called *adaptive applications*) are soft real-time applications that show somewhat higher utility across a greater bandwidth range. This is achieved by adjusting the transmission rate in the presence of congestion, for example, by increasing compression or avoiding transmission of some data blocks.

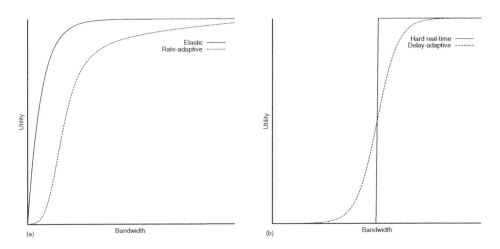

Figure 2.16 Utility functions of several types of applications

Applications that realize a layered multicast congestion control scheme (see Section 2.15) are rate adaptive; because each layer requires a certain amount of bandwidth, a function that increases utility step by step would represent their behaviour more precisely.

2.17.3 Proportional fairness

In order to maximize financial gain, the network should strive to maximize the utility functions of all flows. If these utility functions are logarithmic (i.e. match the criteria for elastic applications), this process can be seen as an optimization (linear programming) problem, the solution of which leads to a rate allocation that satisfies the *proportional fairness* criterion as defined in (Kelly 1997):

> *A vector of rates $\vec{x} = (x_s, s \in S)$ is 'proportionally fair' if it is feasible (..) and if for any other feasible vector $\vec{x}^*$ the aggregate of proportional changes is zero or negative:*

$$\sum_{s \in S} \frac{x_s^* - x_s}{x_s} \leq 0 \qquad (2.3)$$

In our example, a proportionally fair allocation would provide flow 1 with $\frac{c}{3}$ and flows 2 and 3 with $\frac{2}{3}c$ of the available capacity; it seems to be intuitively clear that this rate allocation could lead to a greater financial gain than the rate-maximized vector (*flow*1 = 0, *flow*2 = c, *flow*3 = c), where flow 1 does not obtain any bandwidth and the corresponding user would certainly not be willing to pay anything. Other than max−min fairness, proportional fairness shows a bias against flows with larger RTTs: flows 2 and 3, which use only one resource each, are better off than flow 1, which uses two resources.

Proportional fairness is quite an ideal way of allocating rates, as it is designed to maximize the financial gain for service providers (who may have to install new software in order to realize such a scheme); however, realizing it in an efficient and scalable manner is a difficult problem in practice. Since it was shown that AIMD end-to-end congestion control, which also has the property of giving preference to users with short RTTs, tends to distribute rates according to proportional fairness (Kelly et al. 1998), the same has been said of TCP – but this is in fact only a very rough approximation (Vojnovic et al. 2000). Moreover, it would be more useful to realize a type of fairness that supports heterogeneous utility functions. This, of course, brings about a whole new set of questions. For example, while allocating a rate that is below the limit to a hard real-time application is clearly useless, where should one draw the boundary for a delay-adaptive application? Frameworks that support different utility functions are an active and interesting area of research; examples of related works are (Campbell and Liao 2001) and (Venkitaraman et al. 2002).

2.17.4 TCP friendliness

In the Internet of today, defining fairness follows a more pragmatic approach. Because most of the flows in the network are TCP flows and therefore adhere to the same congestion control rules, and unresponsive flows can cause great harm, it is 'fair' not to push away TCP flows. Therefore, the common definition of fairness in the Internet is called *TCP friendliness*; TCP-friendly flows are also called *TCP compatible* and defined as follows in (Braden et al. 1998):

> *A TCP-compatible flow is responsive to congestion notification, and in steady state it uses no more bandwidth than a conforming TCP running under comparable conditions.*

We will revisit TCP-friendly congestion control and take a look at some of the numerous proposals to realize it in Chapter 4.

2.18 Conclusion

This chapter has given an overview of basic congestion control issues. The goal was to introduce the underlying concepts and principles without bothering you with unnecessary protocol-specific details. The focus in this chapter was on the *why* and *how* as opposed to the *what*. We have seen how congestion comes about and how it can be dissolved, and why mechanisms are designed the way they are.

Congestion control is a world full of problems. Up to now, there is no such thing as the perfect congestion control mechanism; trade-offs are inevitable. By relying solely on implicit feedback, it can be ensured that a mechanism remains scalable and lightweight – but then, assumptions about the environment are made, and one faces problems with heterogeneous environments, various operation modes and so on. On the other hand, if a mechanism is based upon explicit feedback, it may be difficult to keep it scalable. As we have seen, there are lots of additional problems that have to do with incentives, pricing and fairness. Finally, there is a fundamental design problem regarding congestion control and layers – inherently, it is a function dealing with the inner network and should therefore be placed in layer 3, but it seems that nobody has really managed to build a successful solution in this layer so far.

It was another goal of this chapter to illustrate how different modelling methods – vector diagrams, simulation, mathematical modelling using control theory, game theory, and economics – fit into the picture. This information can help you to figure out which tools you ought to apply if you need to solve a congestion control–related problem, and where to look if you must probe further. Some things had to be skipped, however, because they are clearly beyond the scope of this chapter and, in fact, the whole book: for one, an in-depth coverage of the mathematical tools was impossible as it would fill a complete book – it does, and the book may be an interesting reference if you need the tools (Srikant 2004).

Also, we have only very briefly touched upon application data encoding issues in Section 2.15. Layered encoding of other multimedia data, such as audio or video streams, is a large research topic in itself. As a matter of fact, layered schemes are only a subset of the huge body of work about so-called adaptive (multimedia) applications, which are designed to tune the size of application data on the basis of the available capacity. The tricky part is putting it all together: multimedia data often have rate fluctuations of their own. Then, you have a congestion control mechanism underneath that mandates a certain rate. The main goal of an application data encoding scheme is to make the user happy, while the main goal of a congestion control mechanism is to preserve network stability and thereby make everybody else happy. There is a certain conflict of interests here, which is hard to come by.

A good solution is comprehensive, that is, it satisfies all, or at least the most important, constraints. Building it requires knowledge not only of congestion control principles but also of how it is implemented and how researchers envision future realizations. These things are the concern of the chapters to follow.

3

Present technology

This chapter provides an overview of congestion control–related protocols and mechanisms that you may encounter in the Internet of today. Naturally, this situation always changes, as the network itself is in constant flux; some mechanisms are already outdated at the time of writing, but they may still be in use in some places. Others are just on the verge of becoming widely accepted at the time of writing, but they may prevail when you are reading this book. To make things worse, this chapter reflects the author's personal opinion to some degree because it is not always evident what technology is deployed. Sure, there are standards, but sometimes, a bit of interpretation is required in order to tell what is hot from what is not. What is the value of this chapter, then – is it merely a biased snapshot of the current state of the art? The answer is *no* (of course); since the goal was to separate realistic things from futuristic ones, you can be quite sure that whatever we will discuss here *is*, *was* or at least *will be* deployed in large parts of the Internet. This is of course not comprehensive; for instance, some of the things in the next chapter, which the IETF still considers as experimental, are already deployed in some places. The borderline between 'accepted and deployed' and 'experimental' is really quite a fuzzy one.

Our technology overview is split into two major parts. First, we will look at things that happen in Internet end systems such as your own personal computer, which means that first and foremost we have to examine TCP. This protocol makes up the largest portion of this chapter. In TCP, mechanisms for congestion control and reliability are closely interwoven; it is unavoidable to discuss some of these functions together. Then, starting with Section 3.7, we will proceed with things that happen inside the network. Here, we have RED, which may be the only widely deployed active queue management scheme. While our focus is on TCP/IP technology, there is one exception at the end of the chapter: congestion control underneath IP in the context of the ATM Available Bit Rate (ATM ABR) service. ATM ABR rate control is already widely deployed, and it embeds a highly interesting congestion control scheme that is fundamentally different from IP-based networks. Yet, it has lost popularity, and it may appear to be somewhat outdated – thus, we will only discuss it very briefly.

Some things that can be regarded as 'current technology' will not be covered in this chapter: for instance, we will look at QoS schemes, which belong in the 'traffic management' rather than in the 'congestion control' category, in Chapter 5. Also, the immense number

Network Congestion Control: Managing Internet Traffic   Michael Welzl
© 2005 John Wiley & Sons, Ltd

of analyses regarding deficiencies of current technology (e.g. problems with TCP over wireless links) that have been carried out will not be discussed here. Typically, the goal of such research is to first analyse a problem and then solve it – it is therefore more suitable to present the issues together with some of the proposed solutions. Since we are not interested in all the analyses and solution proposals that preceded the standardization of current technology, these things are typically experimental developments for the future, and as such belong in Chapter 4.

3.1   Introducing TCP

TCP is the element of the original Internet (then ARPANET) protocol suite that provides reliable data transmission between two peers across an unreliable IP-based communication channel. A large number of standard documents updated its behaviour over the years; while the original specification did not include any such mechanisms, the protocol now constitutes the foundation of Internet congestion control. TCP has always been a complex beast. As the addition of features did not exactly simplify it, TCP is best understood by roughly following its historic evolvement. We will therefore start with the earliest version and then discuss its gradual refinements. This is not a strict procedure: for example, Figure 3.1 does not show the header exactly as it was originally specified because it would not make much sense to present an outdated header layout. Yet, some documents appear to mark an era in the evolvement of TCP, and we will basically follow these milestones.

Full coverage of the TCP specification is clearly beyond the scope of this book – right now, entering 'TCP' in the RFC editor[1] search engine yields 127 documents. Of course, some of them are irrelevant (obsoleted, which means that they are merely of historic interest), but many are not. The plethora of TCP documents made it so difficult to understand the complete specification that there is now a 'road map' document (Duke et al. 2004) – a guide to reading the RFCs. Its structure helped a great deal in organizing the following sections, which only describe the essential elements of TCP that are relevant for congestion control. These are the mechanisms that influence when a packet is sent into the network; things like the internal state machine for connection maintenance or the checksum calculation

Source Port									Destination Port
Sequence Number									
Acknowledgement Number									
Header Length	Reserved	C W R	E C E	U R G	A C K	P S H	R S T	S Y N	F I N	Window
Checksum									Urgent Pointer
Options (if any)									
Data (if any)									

Figure 3.1 The TCP header

[1] http://www.rfc-editor.org

algorithm are not described because they are less relevant for the purpose of this book. Such details can be found in the related RFCs listed in the 'road map' or in (Stevens 1994).

3.1.1 Basic functions

RFC 793 (Postel 1981b) is the foundation of the TCP standard. It is not the first specification of TCP – as early as 1974, a 'Transmission Control Program' was described in RFC 675 (Cerf et al. 1974) – but from today's perspective, it is the oldest truly important one, as it fully replaced earlier specifications. Reliability and robustness are the main themes in this document. For instance, it contains the following well-known general robustness principle:

> Be conservative in what you do, be liberal in what you accept from others.

Being conservative means to have the protocol do exactly what the specification says and allow nothing else. If the specification does not explicitly prohibit a certain protocol behaviour, this does not mean that implementing it is acceptable. This is the second part of the principle – being 'liberal in what one accepts from others'. If protocol implementation *A*, which contains the special undocumented behaviour, communicates with implementation *B*, which does not, this will lead to failure unless implementation *B* is written in a way that will have it ignore undocumented behaviour. Another important consequence of following this principle is that it facilitates upgrading of the network: when, say, a new value is devised for a field in the TCP header, an old network node that is 'liberal in what it accepts from others' will simply ignore the value instead of rejecting a connection. This way, it is theoretically possible to gradually deploy a new mechanism in the network. As we will see in Section 3.4.9, things are in fact not so easy anymore as firewalls are normally *not* liberal in what they accept from others.

TCP as specified in RFC 793 provides reliable full-duplex[2] transmission of data streams between two peers; the specification does not describe any congestion control features at all. Searching the document for 'congestion' yields the following two statements:

> This document focuses its attention primarily on military computer communication requirements, especially robustness in the presence of communication unreliability and availability in the presence of congestion, but many of these problems are found in the civilian and government sector as well.

> Because segments may be lost due to errors (checksum test failure), or network congestion, TCP uses retransmission (after a timeout) to ensure delivery of every segment.

Interestingly, the latter statement already acknowledged the fact that packets may not only be dropped in the presence of congestion but also when a checksum fails. This ambiguity, which is now the reason for the many known problems with modern TCP implementations across wireless links, was therefore already documented as early as 1981. The statement also mentions *segments* (the TCP data units – the transport layer equivalent of packets),

---

[2]To keep things simple, we will focus on communication that only flows in one direction, with one peer acting as the sender and the other as the receiver.

and it reveals the core element of TCP: reliability by retransmission. TCP segments are carried in IP packets and begin with the header shown in Figure 3.1. The meaning of the fields is as follows:

Source Port/Destination Port (16 bit each): These fields enable differentiating between various applications that run on the same computer; this is a common function in transport layer protocols.

Sequence Number (32 bit): In order to realize retransmission, lost data must be identifiable. This is achieved via this field in TCP; notably, it does not count segments but counts bytes.[3]

Acknowledgement Number (32 bit): ACKs are used to notify the sender of the next sequence number that the receiver is expecting. This field carries the sequence number, and the ACK control flag must be set in order to turn a segment into an ACK.

Header Length (4 bit): This is the number of 32-bit words in the TCP header (used to indicate where the data begin – the length of the TCP header may vary because of its 'Options' field).

Reserved (4 bit): These bits are reserved for future use; they must be zero.[4]

Flags (8 bit): CWR and ECE are used for Explicit Congestion Notification (ECN). The basic idea of ECN was described in Section 2.12.1, and its implementation for TCP/IP will be described in Section 3.4.9. URG is used to signify whether there are urgent data in the segment. If this flag is set, the 'Urgent Pointer' field indicates where in the segment these data begin. This can, for instance, be used to transfer out-of-band interrupt style signalling data for applications like Telnet (Braden 1989). A segment is an ACK if its ACK flag is set.

PSH, the 'push flag', is used for quick message transmission even though the sender buffer is not full. Normally, a TCP sender aggregates data until the *Maximum Segment Size (MSS)*[5] is reached – this can be prevented by setting the push flag. This flag is, for example, necessary at the end of an FTP data stream, where there may just not be enough source data available to fill a whole MSS-sized segment. It should be used with care because small segments are inefficient – the smaller a packet, the larger its relative header overhead, and the more the bandwidth that is wasted.

Finally, RST, SYN and FIN are used to reset, synchronize (i.e. initiate) and finalize a connection, respectively. Their use will briefly be described in the next section.

---

[3]We will sometimes count segments in this chapter for ease of explanation.

[4]This field is a good example of why one should always apply the robustness principle: being 'conservative in what you do' could be interpreted as taking this rule seriously and not 'playing around' with this field. There exists at least one TCP implementation that does not follow this rule – it sets all bits to 1. This had to be taken into account in RFC 3168 (Ramakrishnan et al. 2001), which introduced the CWR and ECE flags and thereby made the Reserved field shrink.

[5]The MSS should equal the size of the largest possible packet that will not be fragmented – the *Maximum Transfer Unit (MTU)* – along the path minus IP and TCP header sizes. The MTU of a path can be determined with a mechanism that is called *Path MTU Discovery (PMTUD)* and specified in RFC 1191 (Mogul and Deering 1990). Sometimes, you might see 'SMSS' in RFCs; this refers to the MSS of the sender (remember that TCP is bidirectional). We will stick with 'MSS' throughout this book for the sake of simplicity.

Window (16 bit): This is the number of bytes (beginning with the one indicated in the acknowledgement field) that the receiver of a data stream is willing to accept. We will discuss this further in Section 3.1.3.

Checksum (16 bit): This is a standard Internet checksum ('Adler-32', the same algorithm that is used in IP (Postel 1981a)) covering the TCP header, the complete data and a so-called 'pseudo header'. The pseudo header is conceptually prefixed to the TCP header; it contains the source and destination addresses and the protocol number from the IP header, a zero field and a virtual 'TCP length' field that is computed on the fly. All these fields together clearly identify a TCP packet.

Urgent Pointer (16 bit): See the description of the URG flag above.

Options (variable): This optional field carries TCP options such as 'MSS' – this can be used by a receiver to inform the sender of the maximum allowed segment size at connection setup. We will discuss some other options in the following sections.

Data (variable): This optional field carries the actual application data. An ACK can carry data in addition to the ACK information (this process is called *piggybacking*), but in practice, this is a somewhat rare case that mainly occurs with interactive applications.

3.1.2 Connection handling

In order to reliably transfer a data stream from one end of the network to the other, TCP first establishes a logical *connection* using a common three-way handshake. This procedure is shown in Figure 3.2; first, Host 1 sends a segment with the SYN flag set in order to initiate a connection. Host 2 replies with a SYN, ACK (both the SYN and ACK flags in the TCP header are set), and finally, Host 1 acknowledges that the connection is now active with an ACK. These first segments of a connection also carry the initial sequence numbers, and they can be used for some additional tasks (e.g. determine the MSS via the option explained above).

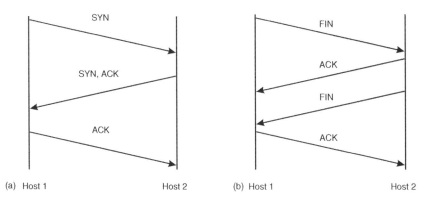

Figure 3.2 Connection setup (a) and teardown (b) procedure in TCP

At the end of the actual data flow, the connection is terminated as shown in Figure 3.2. In contrast to the connection establishment procedure, a four-way handshake is carried out in order to ensure that the connection is correctly terminated from both ends. At this point, we should note that the scenarios depicted in Figure 3.2 are only the simplest, ideal cases. There are many issues here – any of these segments can be lost, or both ends may try to initiate or terminate a connection at the same time. This makes both of these functions (especially teardown) somewhat tricky; a large number of possible failure scenarios are discussed in RFC 793.

A comprehensive description of all the issues related to connection setup and teardown is beyond the scope of this book. Recommendable references are RFC 793 (Postel 1981b), (Stevens 1994) and (Tanenbaum 2003).

3.1.3   Flow control: the sliding window

TCP uses window-based flow control. As explained in Section 2.7, this means that the receiver carries out flow control by granting the sender a certain amount ('window') of data; the sender must not send more than the full window without waiting for acknowledgements at any time. Sliding window protocols require the sender and receiver to keep track of a number of variables; the things that the sender must know about are illustrated in Figure 3.3. In the depicted scenario, ten segments were placed in the buffer for transmission. The receiver advertised a window of six segments; note that all these numbers refer to bytes and not to segments in TCP – thus, the Window field in the header carries a value of $6 * MSS$ in our example. Segments 0 and 1 were transmitted and acknowledged and can be removed from the buffer. Segments 2, 3 and 4 were transmitted but not yet acknowledged, and segments 5, 6 and 7 are ready for transmission because the window spans across these segments.

If an acknowledgement for segment 2 arrives and the advertised window remains the same, the sender can advance ('slide') its window by one segment – then, segment 2 can be discarded and segment 8 becomes ready for transmission. Segment 9 can only be sent when the next acknowledgement arrives. All this remains relatively simple as long as the window size is fixed. In practice, the window can change: for example, the ACK for segment 2 could contain an advertised window of seven segments, which would allow the sender to transmit all the buffered segments (the window *opens*). Similarly, it could contain an advertised window of five segments, which means that the window *closes* – the left edge advances, but the right stays the same.

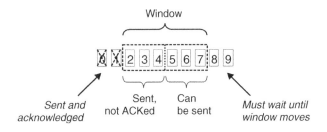

Figure 3.3 The buffer of a TCP sender

What if the advertised window in this ACK is of four segments? Since we cannot skip data, this would mean that the right edge moves to the left – the window *shrinks*, and any segments to the right of this edge theoretically should not have been sent, even if they already were. Shrinking is ugly, and according to RFC 793, it is strongly discouraged but still possible in TCP. RFC 1122 says that a TCP sender must be able to cope with receivers that shrink the window (Braden 1989).

3.1.4 Reliability: timeouts and retransmission

Essentially, TCP achieves reliability by sending a segment, waiting for its ACK, and retransmitting it after a while if the ACK does not arrive. What exactly 'after a while' means is critical: if this value is too small, a sender could spuriously retransmit a segment that is still in flight (or is not yet ACKed), while a value that is too large has the sender wait too long and is therefore inefficient (see Section 2.8 for more details). RFC 793 acknowledges the fact that the *retransmission timeout (RTO)* must be dynamically determined. Without mandating its usage, it lists an example procedure that was obsoleted in RFC 1122 (Braden 1989). We will take a closer look at RTO calculation later.

In the literature, there are several ways to make a protocol carry out error control with retransmissions: the receiver can tell the sender that an expected segment has *not* arrived with a *Negative Acknowledgement (NACK)*, or it can send ACKs that inform the sender of segments that made it. 'ACK 5000' can mean 'resend everything starting from byte number 5000', or it can mean 'resend the segment carrying byte number 5000'. The first case is called *Go-Back-N*, and its main advantage is its simplicity. The second case is called *Selective Repeat*, and it is theoretically more efficient but brings about quite a number of complications.

One major difference between implementations of these two error-control methods is that Selective Repeat requires more sophisticated buffering. For instance, if segments with numbers 1, 3, 4 and 5 reach a Go-Back-N receiver, all it needs to do upon reception of segment no. 3 is request segment no. 2 and wait – there is no need to buffer segments 3, 4 and 5 as they will be retransmitted anyway. With Selective Repeat, the message to the sender is 'just send segment no. 2', segments no. 3, 4 and 5 must be kept, and the data from segment 2 to segment 5 can be given to the receiving application as soon as segment no. 2 arrives. This is obviously more efficient, but is more difficult to implement than Go-Back-N.

Error control in TCP could be seen as some sort of a hybrid: its ACKs have essentially the same cumulative meaning as with Go-Back-N ('retransmit everything from here'), but in TCP, an ACK with a higher number can interrupt the continuous retransmission procedure. The following example from (Kurose and Ross 2004) makes the differences between Go-Back-N and TCP quite obvious: consider a sender that transmits its maximum window of six segments. Let us assume that the window does by no means suffice to saturate the link – it is smaller than half the bandwidth × delay product. This scenario is essentially the same as the one depicted in Figure 2.8, but we do not need to care about congestion for now. Now assume that, say, the second ACK is dropped. With Go-Back-N, the sender would then retransmit the second and the remaining four segments, whereas with TCP, chances are that no retransmission occurs at all. TCP would only retransmit the second segment in case of a timeout – which is cancelled if the third ACK arrives in time. Timeouts lead to actual Go-Back-N behaviour if a whole window of data is dropped.

At this point, it is also notable that the original specification in RFC 793 does not forbid buffering out-of-order segments at the receiver or prescribe when to send an ACK. If, in our example above, segment 2 is successfully retransmitted and arrives at the receiver, where segments 3, 4 and 5 are stored, the receiver could therefore directly answer with ACK 6. We will further elaborate on sending ACKs in the next section.

3.2   TCP window management

3.2.1   Silly window syndrome

RFC 1122 (Braden 1989) made some significant changes to TCP, mainly by mandating the use of mechanisms that were first described elsewhere. Among them are the algorithms for avoiding the *Silly Window Syndrome (SWS)*, an effect that can be caused by small segments. This is how it comes about: as already mentioned, a TCP sender should postpone sending data until an MSS-sized segment can be filled, unless the push flag is set. Waiting is, however, not mandated in RFC 793, and an interactive application might have trouble generating enough data for a whole segment. Imagine, for example, a Telnet user who will only enter a command – and thereby cause traffic – when the result of a previously entered command is visible; if the first command is buffered at the TCP sender side, the user could wait forever. In such a scenario, it might seem efficient to transmit commands or even characters right away. Now imagine that the sender does this, and the small segment that it generates reaches a receiver, which immediately ACKs all segments that come in. The ACK corresponding to the small segment would lead the sender to transmit another small segment, which the receiver would acknowledge, and so on. The small window size stays in the system, and the sender has caused a prevailing problem by sending a single small segment.

3.2.2   SWS avoidance

The SWS is silly indeed, and it is troublesome because one cannot simply forbid sending small segments – we still have the push flag for situations when data must be transmitted right away. The SWS was first described in RFC 813 (Clark 1982) together with three proposed algorithms that conjointly solve the problem: two at the receiver side and one at the sender side. The first algorithm requires the receiver to reduce its advertised window by the size of incoming small segments until the sender can be allowed to transmit a full MSS-sized segment. Then, the advertised window will jump to the new value. Hence, instead of increasing its window in a continuous fashion, the sender will now do so in MSS-sized steps.

Figure 3.4 shows how this algorithm works. The diagrams on the left-hand side represent the sender buffer. An MSS consists of six data blocks; for simplicity, we can say that each of these blocks in the diagram has a size of 1 kB. As in Figure 3.3, the dashed box is the current sender window. The dashed arrow indicates an initial advertised window of 6 kB (one MSS) as a starting point – this arrow could, for instance, represent the last (SYN, ACK) message of the three-way handshake that sets up a connection. Let us assume that the receiver has no problem in accepting data as fast as they arrive and would not normally

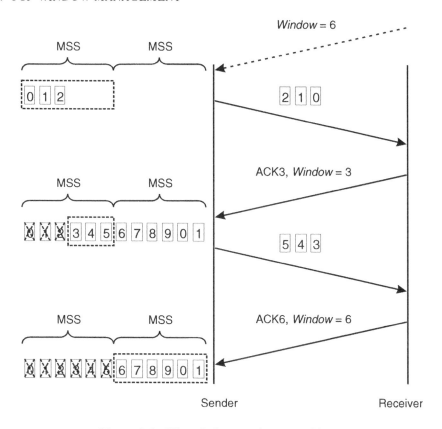

Figure 3.4 Silly window syndrome avoidance

alter the advertised window size, that is, there should always be one MSS in flight. Here is what happens:

- At the beginning, the sender has 3 kB in the buffer but it has a window of a whole MSS available; while we know that it should wait until the MSS can be filled, it does not do so – the first 3 kB are sent right away, potentially leading to SWS.

- The receiver, which implements the SWS avoidance algorithm, reduces the window by advertising only 3 kB instead of 6 kB. Note that it cannot reduce the window further because that would have the right edge of the window move to the left. As mentioned before, this is called *shrinking* the window, and it is discouraged.

- The application has now filled the buffer of the sender. Since the window was only reduced by 3 kB, the sender cannot fill an MSS-sized segment now – all it can do is transmit the 3 kB that it is granted.

- Next, the receiver calculates that the window size would now be zero, and therefore, it can advertise a full window of 6 kB (1 MSS) and hope that the sender fills it this time.

Another possible scenario is that the receiving application cannot accept the data as fast as they arrive for some reason, and consequently the available buffer space is less than an MSS. Obviously, in this case,[6] the receiver should not advertise this small window but rather a window of size zero until the buffer space becomes available again.

3.2.3 Delayed ACKs

If you take close look at Figure 3.4, you may notice that the second solid arrow – the 'ACK 3, Window 3' message from the receiver – is utterly pointless. Without this message, the sender would still be allowed to send the remaining 3 kB. Still, it is a packet in the network that wastes bandwidth and computation resources. Thus, the obvious solution is to delay the ACK somewhat; while RFC 813 suggests that this delay could be fixed to 200 or 300 ms and describes an alternative algorithm to update it dynamically, RFC 1122 simply states that it must not exceed 0.5 s, and that at least one ACK should be sent for every other segment that arrives. This limitation is necessary because excessively delaying an ACK disturbs RTT estimation.

Imagine that in our example the second solid arrow simply does not exist. Then, it becomes evident that delaying an ACK means that additional segments could arrive while the receiver waits for the timer to expire, and that these segments could cause a cumulative ACK, which allows the sender to advance its window in a more reasonable manner. Delayed ACKs have another advantage: they increase the chances for piggybacking, as an application on the receiver side could generate new data while the TCP receiver is waiting for its ACK timer. Also, the fact that ACKs potentially waste bandwidth and computation resources holds true even if we neglect SWS avoidance. After all, who says that generating as much signalling traffic as one ACK per segment is really necessary?

3.2.4 The Nagle algorithm

While the receiver-side algorithms prevent small window sizes and corresponding segments from remaining in the system forever, they do not prevent a sender from transmitting small segments. We have mentioned before that a sender should not do so – but on the other hand, this rule may be a bit too simplistic. Consider our Telnet user again, who, as we learned before, only types a single command, waits for its results to appear on the screen and then goes on to type the next command. In such a case, we are facing a trade-off between interactivity and efficient bandwidth usage. On the one hand, we have the extreme case of a TCP sender that would transmit every single letter that is typed right away, while on the other, we have a TCP sender that always waits until a complete MSS-sized segment is filled (which can take forever in our example). As usual, a reasonable compromise is somewhere in between.

RFC 813 discusses the most obvious (but least efficient) idea of utilizing a timer at the sender side, that is, if data are kept too long, they will be transmitted no matter if they fill an MSS-sized segment or not. Then, a self-clocking algorithm that calculates the ratio of the currently usable window to the advertised window is described; if the ratio is below a certain fraction, one can assume that there is still a significant amount of data in flight and

---

[6]More precisely, the rule says that there should be at least space for one MSS or one-half of the buffer (Stevens 1994).

the window will open some more – thus, the sender should keep the data until the fraction exceeds a certain limit.

The algorithm that was eventually mandated in RFC 1122 is a slightly different one, first described by John Nagle in RFC 896 (Nagle 1984) and termed *Nagle Algorithm*.[7] This algorithm is very simple; here is its description from RFC 1122:

> If there is unacknowledged data, then the sending TCP buffers all user data (regardless of the PSH bit), until the outstanding data has been acknowledged or until the TCP can send a full-sized segment.

This limitation ensures that at least one segment is sent per RTT no matter how much traffic an application generates. Interestingly, the Nagle algorithm even buffers data if the push bit is set; on the other hand, RFC 1122 states that it must be possible for a user to disable this mechanism altogether.

3.3 TCP RTO calculation

As mentioned before (see Sections 3.1.4 and 2.8), it is crucial to have a good estimate of the RTT because it is used for calculating the retransmission timeout (RTO).[8] Have a look at this example procedure from RFC 793:

> Measure the elapsed time between sending a data octet with a particular sequence number and receiving an acknowledgement that covers that sequence number (segments sent do not have to match segments received). This measured elapsed time is the Round Trip Time (RTT). Next compute a Smoothed Round Trip Time (SRTT) as:
>
> $$SRTT = \alpha * SRTT + (1 - \alpha) * RTT$$
>
> and based on this, compute the retransmission timeout (RTO) as:
>
> $$RTO = \min(UBOUND, \max(LBOUND, \beta * SRTT))$$
>
> where *UBOUND* is an upper bound on the timeout (e.g. 1 min), *LBOUND* is a lower bound on the timeout (e.g. 1 s), α is a smoothing factor (e.g. 0.8 to 0.9), and β is a delay variance factor (e.g. 1.3 to 2.0).

In this procedure, a time series model known as an *exponentially.weighted moving average (EWMA)* process is applied. This is known to be a simple and useful predictor when little is known about the nature of incoming data (Makridakis et al. 1997). Moreover, more-sophisticated models often require much more costly operations with non-negligible storage effort. While the EWMA is also found in modern TCP implementations, there are also some significant differences between the procedure above and the current standard.

---

[7]We already encountered this name at the beginning of Chapter 2 because RFC 896 is also the first RFC to mention the term 'congestion collapse'.

[8]Note that there are actually some subtle differences between RTT estimation and RTO calculation (Allman and Paxson 1999): for example, delayed ACKs are an important aspect for RTO calculation but can falsify an RTT estimate. We ignore these details for the sake of simplicity.

3.3.1 Ignoring ACKs from retransmissions

In principle, the first sentence in the procedure described in RFC 793 is still correct: a sample of the current RTT is taken (this is often called the *SampleRTT*) by calculating the time between sending a segment and receiving a corresponding ACK. However, if this is done with each and every segment, a problem might occur because TCP cannot distinguish ACKs that correspond with regular segments from ACKs that correspond with their retransmissions; this is called *retransmission ambiguity* (Karn and Partridge 1987). Retransmitted diagrams were sent an RTO later than the original ones, and their ACKs can therefore drastically falsify RTT estimation. This problem becomes devastating if the ACKed retransmitted segment was not the first, but for instance the second or third try, as the RTO timeout becomes twice as long with each failure.

Doubling the RTO when retransmissions fail is called *exponential backoff*, and its goal is to alleviate congestion somewhat. It is reminiscent of a similar timer-doubling principle that is used for resolving collisions in Ethernet as part of CSMA/CD, and was mandated in RFC 1122 (Braden 1989). Like several other important TCP features, exponential RTO backoff was first described in (Jacobson 1988). It contains the following statement, which explains the underlying reasoning behind this particular feature:

> For a transport endpoint embedded in a network of unknown topology and with an unknown, unknowable and constantly changing population of competing conversations, only one scheme has any hope of working – exponential backoff.

Simple as it may seem, this approach has its merits when designing mechanisms for a network that is as hard to model as the Internet, and it may have guided several other design decisions.[9] In addition to the reasons given in (Chiu and Jain 1989), this also explains why the rate-reduction strategy should be multiplicative (applying it several times leads to exponential decay).

A solution to the retransmission ambiguity problem was presented in (Karn and Partridge 1987) under the name *Karn's Algorithm*, which simply states that ACKs that correspond with retransmitted segments must not be used for RTT estimation. In addition, the algorithm requires the backed-off RTO for this datagram to be used as the RTO for the next datagram – only when it (or a succeeding datagram) is acknowledged without an intervening retransmission will the RTO be recalculated.

3.3.2 Not ignoring ACKs from retransmissions

Ignoring of ACKs from retransmitted segments has a major disadvantage: say a full window of 100 segments is sent, and all these segments are dropped and retransmitted in series. This means that the sender cannot update the RTO for a long time, but the network is in serious trouble and reacting properly is very important. Actually, this is not the only reason why RTO calculation may not be updated frequently enough: RFC 793 does not say anything about how often an RTT sample must be taken. It allows quite a bit of flexibility regarding RTO calculation, as even the procedure described earlier is only provided as a hint. A simplistic implementation could therefore base its calculation on only a single sample per RTT, and this is a severe mistake.

---

[9]For example, RFC 1122 also recommends exponential backoff for 'zero window probing', which is an entirely different issue that we will not discuss here.

Taking less implicit feedback into account than there is available may generally be a bad idea: the more an end system can learn about the network in between, the better. Van Jacobson explained this in a much more precise way in RFC 1323 (Jacobson et al. 1992) by pointing out that RTT estimation is actually a signal processing problem. The frequency of the observed signal is the rate at which packets are sent; if samples of this signal are taken only once per RTT, the signal is sampled at a much lower frequency. This violates Nyquist's criteria and may therefore cause errors in the form of aliasing. This problem is solved in RFC 1323 by the introduction of the *Timestamps option*, which allows a sender to take samples based on (almost) each and every ACK that comes in.

Using the Timestamps option is quite simple. It enables a sender to insert a timestamp in every data segment; this timestamp is reflected in the next ACK by the receiver. Upon receiving an ACK that carries a timestamp, the sender subtracts the timestamp from the current time, which always yields an unambiguous RTT sample. The option is designed to work in both directions at the same time (for full-duplex operation), and only ACKs for new data are taken into account so as to make it impossible for a transmission pause to artificially prolong an RTT estimate. If a receiver delays ACKs, the earliest unacknowledged timestamp that came in must be reflected in the ACK, which means that this behaviour influences RTO calculation. This is necessary in order to prevent spurious retransmissions. The Timestamps option has two notable disadvantages: first, it causes a 12-byte overhead in each data packet, and second, it is known that it is not supported by TCP/IP header compression as specified in RFC 1144 (Jacobson 1990).

3.3.3 Updating RTO calculation

The procedure described in RFC 793 does not work well even if all the samples that are taken are always precise. Before we delve into the details, here are two simple and rather insignificant changes: first, the upper and lower bound values are now known to be inadequate – RFC 1122 states that the lower bound should be measured in fractions of a second and the upper bound should be 240 s. Second, the SRTT calculation line is now typically written as follows (and we will stick with this variant from now on):

$$SRTT = (1 - \alpha) * SRTT + \alpha * RTT \tag{3.1}$$

This is similar to the original version except that α is now $(1 - \alpha)$, that is, a small value is now used for this parameter instead of a large one. RFC 2988 recommends setting α to 1/8 (Paxson and Allman 2000).

The values of α and β play a role in the behaviour of the algorithm: the larger the α, the stronger the influence of new measurements. If the factor β is close to 1, the RTO is efficient in that TCP does not wait unnecessarily long before it retransmits a segment; on the other hand, as already mentioned in Section 2.8, it is generally less harmful to overestimate the RTO than to underestimate it. Clearly, both factors constitute a trade-off that requires careful tuning, and they should reflect environment conditions to some degree. Given the heterogeneity of potential usage scenarios for TCP, one may wonder if fixed values for α and β are good enough.

If, for instance, traffic varies wildly, this can lead to delay fluctuations that are caused by queuing, and it might be better to keep α low and thereby filter such outliers. On the other hand, if frequent and massive delay changes are the result of a moving device, it might be better to have them amply represented in the calculation and choose a larger α. While these statements are highly speculative, some more serious efforts towards adapting these parameters were made: RFC 889 (Mills 1983) describes a variant where α is chosen depending on the relationship between the current RTT measurement and the current value of SRTT. This enhancement, which has the predictor react more swiftly to sudden increases in network delay that stem from queuing, was never really incorporated in TCP – the most-recent specification of RTO estimation, RFC 2988, still uses a fixed value. A measurement study indicates that its impact is actually minor (Allman and Paxson 1999), and that the minimum RTO value is a much more important parameter. It must be set to 1 s accord-ing to RFC 2990, which says that this a 'conservative approach, while at the same time acknowledging that at some future point, research may show that a smaller minimum RTO is acceptable or superior'.

In order to understand the meaning of β, remember that we want to be on the safe side – the calculated RTO should always be more than an RTT because it is the most-important goal to avoid ambiguous retransmits. If the RTTs are relatively stable, this means that having a little more than an average RTT might be safe enough. On the other hand, if RTT fluctuation is severe, it might be better to have some overhead – something like, say, twice the estimated RTT might be more appropriate than just using the estimated RTT as it is in such a scenario. This factor of cautiousness is represented by β in the RFC 793 description; its value should depend on the magnitude of fluctuations in the network.

A major change was made to this idea of a fixed β in (Jacobson 1988): since it is known from queuing theory that the RTT and its variation increase quickly with load, simply using the recommended value of 2 does not suffice to cover realistic conditions. The paper gives a concrete example of 75% capacity usage, leading to an RTT variation factor of sixteen, and notes that $\beta = 2$ can adapt to loads of at most 30%. On the other hand, constantly using a fixed value that can accommodate such high traffic occurrences would clearly be inefficient. It is therefore better to have β depend on the variation instead; in an appendix of his paper, Jacobson proposes using the mean deviation instead of the variation for ease of computation. Then, he goes on to describe a calculation method that is optimized to compensate for adverse effects from limited clock granularity as well as computation speed.

The very algorithm described in (Jacobson 1988) can be found in the kernel source code of the Linux machine that I used to write this book. It might seem that the speed of calculation may have become less important over the years; while it is probably true that it is not as important as it used to be, it is still not totally irrelevant, given the diversity of appliances that we expect to run a TCP/IP stack nowadays. Neglecting a detail that is related to clock granularity, the final equations that incorporate the variation σ (or actually its approximation via the mean deviation) in RFC 2988 are

$$\sigma = (1 - \beta) * \sigma + \beta * [SRTT - RTT] \tag{3.2}$$

$$RTO = SRTT + 4 * \sigma \tag{3.3}$$

where $[SRTT - RTT]$ is the prediction error and β is 1/4. Note that setting β to 1/4 and α to 1/8 means that the variation will more rapidly react to fluctuations than the RTT estimate, and adding four times[10] the variation to the SRTT for RTO calculation was done in order to

---

[10]The original version of (Jacobson 1988) suggested calculating RTO as $SRTT + 2 * \sigma$; practical experience led Jacobson to change this in a slightly revised version of the paper.

avoid adverse interactions with two other algorithms that he described in the same paper: *slow start* and *congestion avoidance*. In the following section, we will see how they work.

3.4 TCP congestion control and reliability

By describing two methods that limit the amount of data that TCP sends into the network on the basis of end-to-end feedback, Van Jacobson added congestion control functionality to TCP (Jacobson 1988). This could perhaps be seen as the milestone that started off all the Internet-oriented research in this area, but it does not mean that it was the first such work: the paper has a reference to a notable predecessor – *CUTE* (Jain 1986) – which shows many similarities. The mechanisms by Van Jacobson were refined over the years, and some of these updates did not directly influence the congestion control behaviour but only relate to reliability; yet, they are important pieces of the puzzle, which shows the dynamics of modern TCP stacks. Let us now build this puzzle from scratch, starting with the first and fundamental pieces.

We already encountered the 'conservation of packets principle' in Section 2.6 (Page 19). The idea is to stabilize the system by refraining from sending a new packet into the network until an old packet leaves. According to Jacobson, there are only three ways for this principle to fail:

1. A sender injects a new packet before an old packet has exited.

2. The connection does not reach equilibrium.

3. The equilibrium cannot be reached because of resource limits along the path.

The first failure means that the RTO timer expires too early, and it can be taken care of by implementing a good RTO calculation scheme. We discussed this in the previous section. The solution to the second problem is the slow start algorithm, and the congestion avoidance algorithm solves the third problem. Combined with the updated RTO calculation procedure, these three TCP additions in (Jacobson 1988) indeed managed to stabilize the Internet – this was the answer to the global congestion collapse phenomenon that we discussed at the beginning of this book.

3.4.1 Slow start and congestion avoidance

Slow start was designed to start the 'ACK clock' and reach a reasonable rate fast (we will soon see what a 'reasonable rate' is). It works as follows: in addition to the window already maintained by the sender, there is now a so-called *congestion window (cwnd)* also, which further limits the amount of data that can be sent. In order to keep the flow control functionality active, the sender must restrain its window to the minimum of the advertised window and *cwnd*. The congestion window is initialized with one[11] segment and increased by one segment for each ACK that arrives. Expiry of the RTO timer (which, since we now have a reasonable calculation method, can be assumed to mean that a segment was lost) is taken as an implicit congestion feedback signal, and it causes *cwnd* to be reset to one

---

[11]Actually, the initial window is slightly more than one, as we will see in Section 3.4.4 – but let us keep things simple and assume that it is one for now.

segment. Note that this method is prone to all the pitfalls of implicit feedback that we have discussed in the previous chapter.

The name 'slow start' was chosen not because the procedure itself is slow, but because, other than existing TCP implementations of the time, it starts with only one segment (on a side note, the algorithm was originally called *soft start* and renamed upon a message that John Nagle sent to the IETF mailing list (Jacobson 1988)). Slow start is in fact exponentially fast: one segment is sent, and one ACK is received – *cwnd* is increased by one segment. Now, two segments can be sent, which causes two ACKs. For each of these two ACKs, *cwnd* is increased by one such that *cwnd* now allows four segments to be sent, and so on.

The second algorithm, 'congestion avoidance', is a pure AIMD mechanism (see Section 2.5.1 on Page 16 for further details). Once again, we have a congestion window that restrains the sender in addition to the advertised window. However, instead of increasing *cwnd* by one for each ACK, this algorithm usually increases it as follows:

$$cwnd = cwnd + MSS * MSS/cwnd \qquad (3.4)$$

This means that the window will be increased by at most one segment per RTT; it is the 'Additive Increase' part of the algorithm. Note that we are (correctly) counting in bytes here, while we are mostly using segments throughout the rest of the book for the sake of simplicity.

While RFC 2581 only mentions that Equation 3.4 provides an 'acceptable approximation', it is very common to state that this equation has the rate increase by *exactly* one segment per RTT. This is incorrect, as pointed out by Anil Agarwal in a message sent to the end2end-interest mailing list in January 2005. Let us go through the previous example of starting with a single segment again (i.e. *cwnd = MSS*) to see how the error occurs, and let us assume that MSS equals 1000 for now.

One segment is sent,[12] one ACK is received, and *cwnd* is increased by $MSS * MSS/cwnd = 1000$. Now, two segments can be sent, which causes two ACKs. If *cwnd* would be fixed throughout an RTT, it would be increased by $1000 * 1000/2000 = 500$ for each of these ACKs, leading to a total increase of exactly one MSS per RTT. Unfortunately, this is not the case: when the first ACK comes in, the sender already increases *cwnd* by $MSS * MSS/cwnd$, which means that its new value is 2500. When the second ACK arrives, *cwnd* is increased by $1000 * 1000/2500 = 400$, yielding a total *cwnd* of 2900 instead of 3000. The sender cannot send three but can send only two segments, leading to at most two ACKs, which further prevents *cwnd* from growing as fast as it should.

This effect is probably negligible if the sending rate is high and ACKs are evenly spaced, as *cwnd* is likely to be increased beyond 3000 when the next ACK arrives in our example; this would cause another segment to be sent soon. It might be a bit more important when *cwnd* is relatively small (e.g. right after slow start), but since this does not change the basic underlying AIMD behaviour, it is, in general, a minor issue; this appears to be the reason why the IETF has not changed it yet. Also, while increasing by exactly one segment per RTT is the officially recommended behaviour, it may in fact be slightly too aggressive. We will give this thought further consideration in Section 3.4.3.

The exponential increase of slow start and additive increase of congestion avoidance are depicted in Figure 3.5; note that starting with only one segment and increasing by

---

[12]Starting congestion avoidance with only one segment may be somewhat unrealistic, but it simplifies our explanation.

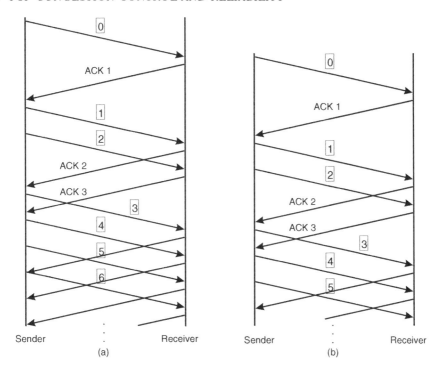

Figure 3.5 Slow start (a) and congestion avoidance (b)

exactly one segment per RTT in congestion avoidance as in this diagram is an unrealistic simplification. Theoretically, the 'Multiplicative Decrease' part of the congestion avoidance algorithm comes into play when the RTO timer expires: this is taken as a sign of congestion, and *cwnd* is halved. Just like the additive increase strategy, this differs substantially from slow start – yet, both algorithms have their justification and should somehow be included in TCP.

3.4.2 Combining the algorithms

In order to realize both slow start and congestion avoidance, the two algorithms were merged into a single congestion control mechanism, which is implemented at the sender as follows:

- Keep the *cwnd* variable (initialized to one segment) and a threshold size variable by the name of *ssthresh*. The latter variable, which may be arbitrarily high at the beginning according to RFC 2581 (Allman et al. 1999b) but is often set to 64 kB, is used to switch between the two algorithms.

- Always limit the amount of segments that are sent with the minimum of the advertised window and *cwnd*.

- Upon reception of an ACK, increase *cwnd* by one segment if it is smaller than *ssthresh*; otherwise increase it by $MSS * MSS/cwnd$.

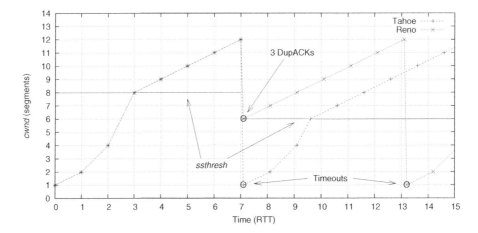

Figure 3.6 Evolution of *cwnd* with TCP Tahoe and TCP Reno

- Whenever the RTO timer expires, set *cwnd* to one segment and *ssthresh* to half the current window size (the amount of data in flight).

Another way of saying this is that the sender is in slow start mode until the threshold is reached; then, it is in congestion avoidance mode until packet loss is detected and it switches back to slow start mode again.

The 'Tahoe' line in Figure 3.6 shows slow start and congestion avoidance interaction (for now, ignore the other line). The name *Tahoe* is worth explaining: for some reason, it has become common to use names of places for different TCP versions. Tahoe is located in the far east of California, and it is well worth visiting – Lake Tahoe is very beautiful and impressively large, and the surrounding area is great for hiking.[13] Usually, each of these versions comes with a major congestion control change. TCP Tahoe is TCP as it was specified in RFC 1122 – essentially, this means RFC 793 plus everything else that we have discussed so far except the Timestamps option (the algorithms for SWS avoidance, updated RTO calculation and slow start/congestion avoidance algorithms). TCP Tahoe is also the BSD Network Release 1.0 in 4.3 BSD Unix (Peterson and Davie 2003).

Note that there are some subtleties that render Figure 3.6 somewhat imprecise: first, as *cwnd* reaches *ssthresh* after approximately 9.5 RTTs, the sender seems to go right into congestion avoidance mode. This is correct according to (Jacobson 1988), which mandated that slow start is only used if *cwnd* is *smaller* than *ssthresh*. In 1997, however, RFC 2001 (Stevens 1997) specified that a sender is in slow start if *cwnd* is smaller *or equal to ssthresh*, whereas the most-recent specification (RFC 2581 (Allman et al. 1999b)) says that the sender can use either slow start or congestion avoidance if *cwnd* is equal to *ssthresh*.

The second issue is that the congestion window reductions after 7 and 13 RTTs happen as soon as the sender receives an ACK – how long the change really takes depends on the ACK behaviour of the receiver. After nine RTTs, *cwnd* equals four, and the sender is in slow start mode and keeps increasing its window by one segment for every ACK

---

[13]As a congestion control enthusiast, I had to go there, and it was also the first time I ever saw an American squirrel up close, which, unlike our Austrian squirrels here, has no bushy tail and does not jump from tree to tree.

that arrives. After two out of the four expected ACKs, it reaches *ssthresh* and continues in congestion avoidance mode – this process takes less than one full RTT, which is indicated by the line reaching *ssthresh* earlier. Once again, the exact duration depends on the ACK behaviour of the receiver. Third, we have already seen that increasing the rate by exactly one segment per RTT in congestion avoidance mode is desirable but it is not what all TCP implementations do.

3.4.3 Design rationales and deployment considerations

Here are some of the reasons behind the slow start and congestion avoidance design choices as per (Jacobson 1988):

- Additively increasing and multiplicatively decreasing was identified as a reasonable control strategy in (Chiu and Jain 1989).

- CUTE used 7/8 as the decrease factor (the value that the rate is multiplied with when congestion occurs). A reason to use 1/2 for TCP instead was that one should use a window size that is known to work – and during slow start, it is clear that half the current window size just worked well. Jacobson lists two reasons for halving the window when packet loss occurred during congestion avoidance: first, if packet loss occurred during congestion avoidance, it is probable that there are now exactly two instead of one flow in the network. The new flow that has entered the network is now consuming half of the available bandwidth, which means that one should reduce its window by half. Second, if there are more than two flows, halving the window is conservative, and being conservative in the presence of a lot of other traffic is probably a good idea.

- Jacobson states in (Jacobson 1988) that the 1-packet-per-RTT increase has less justification than the factor 1/2 decrease and is, in fact, 'almost certainly too large'. In particular, he says:

  > If the gateways are fixed so they start dropping packets when the queue gets pushed past the knee, our increment will be much too aggressive and should be dropped by about a factor of four.

- As mentioned before, the intention of slow start is to start the ACK clock and reach a reasonable rate (*ssthresh*) fast in a totally unknown environment (as, for example, at the very beginning of the communication).

Quite a number of years have passed since (Jacobson 1988) was published. For instance, one may question the validity of the first statement to justify a decrease factor of 1/2 given the length of end-to-end paths and amount of background traffic in the Internet of today. The second one is, however, still correct; the fact that TCP has survived the immense growth of the Internet can perhaps be attributed to this prudence behind its design.

As for the additive increase factor, one could perhaps regard active queue management schemes like RED as such a fix that 'drops packets when the queue gets pushed past the knee'. Therefore, one can also question whether it is a good idea to constantly increase the rate by a fixed value in modern networks. Jacobson also mentions the idea of a second-order control loop to adaptively determine the appropriate increment to use for a path. This shows that he did not regard this fixed way of incrementing the window size as immovable. It is especially interesting to see that Van Jacobson even explicitly stated this in his seminal 'Congestion Avoidance and Control' paper, which is frequently used as a means to defend the mechanisms therein, which some might call the 'holy grail' of Internet congestion control.

On a side note, increasing by significantly less[14] than one packet per RTT is unlikely to be reasonable for the Internet of today unless it is combined with a method to emulate the average aggressiveness of legacy TCP. This is an incentive issue resembling the tragedy of the commons (see Section 2.16 on Page 44) – the question on the table is: why would I want to install a better TCP implementation if it *degrades* my own network throughput at first, until enough other users installed it? One could actually take this thinking a step further and question why slow start and congestion avoidance made it into our protocol stacks in the first place; why did network administrators install it, when it only reduced their own rate at first and brought a benefit provided that enough others installed it, too? It could have to do with the attitude in the Internet community at that time, but there may also be a different explanation: the operating system patch that contained slow start and congestion avoidance also contained the change to the RTO estimation. This latter change, which replaced the fixed value of β with a variation calculation, was reported to lead to an immense performance gain in some scenarios (RFC 1122 mentions one case where a vendor saw link utilization jump from 10 to 90%).

A patch can, of course, be altered. Code can be changed. While it might have been trust in the quality of Jacobson's code that prevented administrators from altering it when it came out, it is hard to tell what now prevents script kiddies from making the TCP implementation in their own operating systems more aggressive. Is it the sheer complexity of the code, or simply lack of incentives to do so (because *taking* (receiving, or *downloading*) is usually more important to them than *giving* (sending, or *uploading*))? In the latter case, there are still options to attain higher throughput by changing the receiver side only (see Section 3.5). Are these possibilities just not known enough – or are some script kiddies out there already fiddling with their TCP code, and we are not aware of it? It is hard to find an answer to these questions. We will further elaborate on these and related issues in Chapter 6; for now, let us continue with technical TCP specifics.

3.4.4 Interactions with other window-management algorithms

In Section 3.2.3, we learned some reasons why a receiver should delay its ACK, and that RFC 1122 mandates not waiting longer than 0.5 s and recommends sending at least one ACK for every other segment that arrives. Under normal circumstances, this means that exactly one ACK is sent for every other segment. This is at odds with the congestion avoidance algorithm, which has the sender increase *cwnd* by $MSS * MSS/cwnd$ for every ACK that arrives. Consider the following example: *cwnd* is 10, and 10 segments are sent within an RTT. If these 10 segments cause 10 ACKs, *cwnd* is additively increased 10 times, which means that it is eventually increased by at most one MSS at the end of this RTT. If, however, the receiver sends only one ACK for every other segment that arrives,

---

[14]As we will see in the next chapter, researchers actually put quite a bit of effort into the idea of increasing by *more* than one segment per RTT, and there are good reasons to do so; see Section 4.6.1.

cwnd is increased by at most MSS/2 during this RTT, and the result is overly conservative behaviour during the congestion avoidance phase.

Interestingly, the congestion avoidance increase rule can also be too aggressive. In Section 3.2.2, we have seen that, if the sender transmits less than an MSS (i.e. the Nagle algorithm is disabled), the receiver ACKs small amounts of data until a full MSS is reached because it cannot shrink the window. These ACKs can sometimes be eliminated by a delayed cumulative ACK, but this requires enough data to reach the receiver before the timer runs out; moreover, delaying ACKs is not mandatory, and some implementations might not do it. It can therefore happen that ACKs that acknowledge the reception of less than a full MSS-sized segment reach the sender, where the rate is updated for each ACK received regardless of how many bytes are ACKed. So far, there is no widely deployed solution to this problem. A reasonable approach that can be implemented in accordance with the most-recent congestion control specification (RFC 2581) is *appropriate byte counting (ABC)*, which we will discuss in the next chapter (Section 4.1.1) because it is still an experimental proposal.

Delayed ACKs are also a poor match for slow start because it begins by transmitting only one segment and waits for an ACK before the next segment is sent. If a receiver always delays its ACK, the delay between transmitting the first segment of a connection and arrival of its corresponding ACK will therefore be significantly increased because the receiver waits for the DelACK timer to expire. Often, this timer is set to 200 ms, but, as mentioned before, RFC 1122 even allows an upper limit of 0.5 s. This constant delay overhead can become problematic when connections are as short as HTTP requests from a web browser; this was one of the reasons to allow starting with more than just a single segment. RFC 3390 (Allman et al. 2002) specifies the upper bound for the *Initial Window (IW)* as

$$IW = \min(4 * MSS, \max(2 * MSS, 4380 \text{ bytes})) \tag{3.5}$$

There are also positive effects from interactions between congestion control and the other window-management algorithms in TCP: theoretically, a sender could actually change its rate (not just the internal *cwnd* variable) more frequently than once per RTT – it could increase it in $1/cwnd$ steps with each incoming ACK by sending smaller datagrams. Then, it does not exhibit the desired behaviour of adding exactly one segment every RTT and nothing in between RTTs. This, however, would require disabling the Nagle algorithm, which is possible but discouraged because it can lead to SWS.

3.4.5 Fast retransmit and fast recovery

On the 30th of April 1990, Van Jacobson sent a message to the IRTF end2end-interest mailing list. It contained two more sender-side algorithms, which significantly refine congestion control in TCP while staying interoperable with existing receiver implementations. They were mainly intended as a solution for poor performance across long fat pipes (links with a large bandwidth × delay product), where one can expect to see the largest gain from applying them, but since they work well in all kinds of situations and also do not sufficiently solve the problems encountered with these links, the new algorithms are regarded as a general enhancement. The idea is to use a number of so-called *duplicate ACKs (DupACKs)* as an indication of packet loss. If a sender transmits segments 1, 2, 3, 4 and 5 and only segments 1, 3, 4 and 5 make it to the other end, the receiver will typically respond to segment 1 with an 'ACK 2' ('I expect segment 2 now') and send three more such ACKs (*duplicate* ACKs) in response to segments 3, 4 and 5; ACKing such out-of-order segments was already mandated in RFC 1122 in anticipation of this feature. These ACKs should not be delayed according to RFC 2581.

At the sender, the reception of duplicate ACKs can indicate that a segment was lost. Since it can also indicates that packets were reordered or duplicated within the network, it is better to wait for a number of consecutive DupACKs to arrive before assuming loss; for this, Jacobson described a 'consecutive duplicates' threshold, which was later replaced with a fixed value (RFC 2581 specifies three DupACKs, that is, four identical ACKs without the arrival of any intervening segments). The first mechanism that is triggered by this loss-detection scheme is *fast retransmit*, which simply lets the receiver retransmit the segment that was requested numerous times without waiting for the RTO timer to expire.

From a congestion control perspective, the more-interesting algorithm is *fast recovery*: since a receiver will only generate ACKs in response to incoming segments, duplicate ACK do not only have the potential to signify bad news (loss) – receiving a DupACK also means that an out-of-order segment has arrived at the receiver (good news). In his email, Jacobson pointed out that if the 'consecutive duplicates' threshold (the number of DupACKs the sender is waiting for) is small compared to the bandwidth × delay product, loss will be detected while the 'pipe' is almost full. He gave the following example: if the threshold is three segments (the standard value) and the bandwidth × delay product is around 24 kb or 16 packets with the common size of 1500 byte each, at least[15] 75% of the packets needed for ACK clocking are in transit when fast retransmit detects a loss. Therefore, the 'ACK clock' does not need to be restarted by switching to slow start mode – just like *ssthresh*, *cwnd* is directly set to half the current amount of data in flight.

This behaviour is shown by the 'Reno' line in Figure 3.6. While the Tahoe release of the BSD TCP code already contained fast retransmit, fast recovery only made it into a release, which was called *Reno*. Geographically, Reno is close to Tahoe (albeit in Nevada), but, unlike Tahoe, it is probably not worth visiting. I still remember the face of the man at the Reno Travelodge check-in desk, who raised his eyebrows when I asked him whether he can recommend a jazz club nearby, and replied: 'In this town, sir?' He went on to explain that Reno has nothing but casinos and a group of kayak enthusiasts, and nobody would probably live there if given a choice. While this is, of course, an extremely biased description, it is probably safe to say that Reno, the 'biggest little city in the world', is a downscaled version of Vegas, which, as we will see in the next chapter, is also a TCP version.

Fast recovery is actually a little more sophisticated: since each DupACK indicates that a segment has left the network, an additional segment can be sent to take its place for every DupACK that arrives at the sender. Therefore, *cwnd* is not set to *ssthresh* but to *ssthresh* + 3 * *MSS* when three DupACKs have arrived. Here is how RFC 2581 specifies the combined implementation of fast retransmit and fast recovery:

1. When the third duplicate ACK is received, set *ssthresh* to no more than half the amount of outstanding data in the network (i.e. at most *cwnd*/2), but at least to 2 * MSS.

---

[15]It is probably a little more than 75% because the pipe is 'overfull' (i.e. some packets are stored in queues) when congestion sets in.

2. Retransmit the lost segment and set *cwnd* to *ssthresh* plus 3 * *MSS*. This artificially 'inflates' the congestion window by the number of segments (three) that have left the network and which the receiver has buffered.

3. For each additional duplicate ACK received, increment *cwnd* by *MSS*. This artificially inflates the congestion window in order to reflect the additional segment that has left the network.

4. Transmit a segment, if allowed by the new value of *cwnd* and the receiver's advertised window.

5. When the next ACK arrives that acknowledges new data, set *cwnd* to *ssthresh* (the value set in step one). This is termed *deflating* the window.

   This ACK should be the acknowledgement elicited by the retransmission from step one, one RTT after the retransmission (though it may arrive sooner in the presence of significant out- of-order delivery of data segments at the receiver). Additionally, this ACK should acknowledge all the intermediate segments sent between the lost segment and the receipt of the third duplicate ACK, if none of these were lost.

Note that it is possible that the algorithm never reaches step five because this ACK does not arrive; in this case, a retransmission timeout occurs, which takes the sender out of fast retransmit/fast recovery mode and brings it into slow start mode.

3.4.6 Multiple losses from a single window

The fast retransmit/fast recovery algorithm is known to show problems if numerous segments are dropped from a single window of data. While timeout-initiated retransmissions essentially have the sender restart from the first unacknowledged segment unless an ACK with a higher number interrupts the process, the algorithm described above retransmits exactly *one* segment in response to three DupACKs. This also means that it will not retransmit more than one segment per RTT. Moreover, the first regular ACK following the DupACKs (the ACK mentioned in step five) implicitly conveys some unused information. An example will show how this happens.

Consider the scenario depicted in Figure 3.7. Let us assume that three DupACKs reliably indicate that a segment was lost and let us neglect the possibility of packet duplication or reordering in the network. When the third duplicate ACK (requesting segment 1) is received, the sender knows that segment 1 was lost and three of the other transmitted segments made it. Hence, fast retransmit/fast recovery sets in, which means that segment 1 is retransmitted, *ssthresh* and *cwnd* are updated and *cwnd* is inflated by 3 * MSS. Note that the sender does not know which ones out of the four segments that were sent after segment 1 reached the receiver. In particular, it is impossible to deduce the loss of segment 3 from these DupACKs – the sender cannot tell the depicted case from a scenario where segments 2, 3 and 4 caused the ACKs for segment 1 and the ACK caused by segment 5 is still outstanding.

The inflated *cwnd* now allows the sender to keep transmitting segments as duplicate ACKs arrive, and the first segment that is sent upon reception of the third DupACK is segment 1. Since all subsequent segments will only generate further DupACKs, it takes one RTT until the next regular ACK that conveys some information regarding which segments

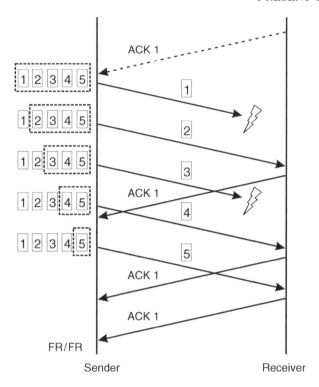

Figure 3.7 A sequence of events leading to Fast Retransmit/Fast Recovery

actually made it to the receiver arrives. This is the ACK that brings the sender out of fast retransmit/fast recovery mode, and it is caused by the retransmitted segment 1. While this ACK would ideally acknowledge the reception of segments 2 to 5, it will be an 'ACK 3' in the scenario shown in Figure 3.7. This ACK, which covers some but not all of the segments that were sent before entering fast retransmit/fast recovery, is called a *partial ACK*.

Segment 3 will be retransmitted if another three DupACKs arrive and fast retransmit/fast recovery is triggered again. The requirement for three incoming DupACKs in response to a single lost segment is problematic at this point. Consider what happens if the advertised window is 10 segments, *cwnd* is large enough to transmit all of them, and every other segment in flight is dropped. For all these segments to be recovered using fast retransmit/fast recovery, a total of 15 DupACKs would have to arrive. Since DupACKs are generated only when segments arrive at the receiver, the sender will not be able to send enough segments and reach a point where it waits in vain for DupACKs to arrive. Then, the RTO timer will expire, which means that the sender will enter slow start mode.

This is undesirable because it renders the connection unnecessarily inefficient: expiry of the RTO timer should normally indicate that the 'pipe' has emptied, but this is not the case here – it is just not as full as it would be if only a single segment was dropped from the window. The problem is aggravated by the fact that *ssthresh* is probably very small (e.g. if it *was* possible to enter fast retransmit/fast recovery several times in a row as described in (Floyd 1994), *ssthresh* would be halved each time). Researchers have put significant effort into the development of methods to avoid unnecessary timeouts – the RTO timer is generally seen as a back-up mechanism that is invoked only when everything else fails.

RFC 3042 (Allman et al. 2001) recommends a very simple method to reduce the chance of RTO timeouts: instead of merely waiting for all the three DupACKs, the sender is allowed to send a *new* segment for each of the first two DupACKs, provided that this is allowed by the advertised window of the receiver. This method, which is called *limited transmit*, can enable the receiver to send two more DupACKs than it would normally do, thereby increasing the chance for the necessary three DupACKs to arrive. Implementing limited transmit is particularly worthwhile when the window is very small and hence the chance of sending enough segments for the receiver to generate the three DupACKs is small, too.

3.4.7 NewReno

One solution to the problem of TCP with multiple drops from a single window is described in (Hoe 1995) and (Hoe 1996). The recommended change to fast retransmit/fast recovery is a very small one, and it is specified in RFC 2582 (Floyd and Henderson 1999) as follows:

- In step five of the original algorithm, the highest sequence number transmitted is stored in a variable called *recover*. This value is later used to distinguish between regular ('full') ACKs and partial ACKs.

- In step five of the original algorithm, the sender distinguishes between a partial ACK and a regular full ACK by checking whether all the data up to and including 'recover' are acknowledged.

  - If the ACK is a partial ACK, the first unacknowledged segment is transmitted (segment 3 in the scenario of Figure 3.7), *cwnd* is 'partially deflated' by the amount of new data acknowledged plus one segment, and a segment is sent if permitted by the new value of *cwnd*. The goal of this procedure is to ensure that approximately *ssthresh* bytes are in flight when fast recovery ends. Then, the sender stays in fast recovery mode (i.e. it goes back to step three of the original procedure).

  - If the ACK is a full ACK, *cwnd* can be set to *ssthresh*. Since this means that the amount of data in flight can now be much less than what the new congestion window allows, a sender must additionally take precautions against generating a sudden burst of data. Alternatively, the sender can set *cwnd* to the minimum of *ssthresh* and the amount of data in flight plus one MSS. In any case, the window is deflated and fast recovery ends.

This TCP variant is called *NewReno*. While the essence of the idea – utilizing the additional information conveyed by a partial ACK – remains, the specification in RFC 2582 is slightly different than the algorithm described in (Hoe 1996). For example, Janey Hoe suggested sending a segment for every two DupACKs that arrive in order to keep the ACK clock in motion, but this idea was abandoned in the specification. Also, instead of retransmitting only a single segment in response to a partial ACK, it was originally envisioned to retransmit lost segments from a single window using the slow start algorithm. This is a more aggressive method, which is able to recover faster from a large number of losses that belong to a single window, but it also has a greater chance of unnecessarily retransmitting a segment.

There are several additional issues related to NewReno, including the questions of when to reset the timer and how to avoid multiple fast retransmits. Since NewReno stays in fast retransmit/fast recovery mode until either a full ACK arrives or the RTO timer expires, only the latter event can cause multiple fast retransmits. In this case, however, it is generally safe to assume that the pipe is empty, and chances are that interpreting three DupACKs as an indication of packet loss is misleading. This effect, which is called *false fast retransmits* in (Hoe 1996), can occur as follows: the receiver has some segments from the previous window in its cache and, for some reason, the sender begins to retransmit them in sequence. Say, the segments cached at the receiver are numbered 4, 5, 6 and 7, and the sender window is large enough to retransmit the three segments 4, 5 and 6. Each of them will cause an 'ACK 8', which looks just like a loss notification.

Here, the problem is that fast retransmit will be entered multiple times even though this is based upon information from a single window. RFC 2582 specifies the following simple fix to avoid this: when a timeout occurs, the highest transmitted sequence number is stored in a variable (which is set to the initial sequence number at the beginning). Before starting the fast retransmission procedure, compare the stored sequence number with the sequence number in the incoming DupACKs. If the stored number is greater than the ACKed number, the ACKs refer to data from the same window, and fast retransmission should not be restarted. RFC 2582 was obsoleted by RFC 3782 (Floyd et al. 2004). This document contains an update of this procedure, which makes intelligent use of the 'recover' variable both for its original purpose and for detection of whether fast retransmit should be restarted.

Even with this fix, NewReno suffers from the fundamental problem that the sender cannot distinguish between a DupACK that is caused by unnecessary retransmissions and a DupACK that correctly indicates a lost or delayed segment. Moreover, the sender cannot tell which segment triggered a DupACK. This causes further problems. For instance, RFC 3782 states that DupACKs with a sequence number that is smaller than the stored maximum can occur if a retransmitted segment is itself lost. Then, it would be better to restart the fast retransmit/fast recovery procedure, but the algorithm does not do this. RFC 3782 describes two different heuristics to detect whether three DupACKs that do not acknowledge more than the stored variable are caused by unnecessary retransmissions or not.

To conclude, we can say that NewReno is a fix based on incomplete information, and it can only alleviate the resulting negative effects to a certain degree. One can think of several ways to make this algorithm more sophisticated, and it is often possible to find scenarios where such small changes are beneficial. Then, the main question on the table is whether such changes are generally useful enough and if they are worth the effort of updating the code in TCP/IP stacks. In any case, it is an inevitable consequence of the incoming incomplete information that a sender only has two choices if multiple segments are dropped from a single window of data: it can either retransmit at most one segment per RTT or take chances to unnecessarily retransmit segments that already made it to the other end (Fall and Floyd 1996). An architecturally better solution would tackle the underlying problem of misleading and incomplete information reaching the sender – but this means that the receiver implementation must also change.

Kind = 5	Length
Left Edge of 1st Block	
Right Edge of 1st Block	
...	
Left Edge of nth Block	
Right Edge of nth Block	

Figure 3.8 The TCP SACK option format

3.4.8 Selective Acknowledgements (SACK)

Enter SACK. Here, a receiver uses a new ACK format that enables it to explicitly say something like 'I received segments 1 to 3, 5, 7, and 9 to 13' (which means that the ones in between are missing) in one go. The sender uses this information to directly retransmit only the needed segments – this moves TCP a significant step closer to being a pure Selective Repeat protocol. Since the above segments can be expected to stem from the same window, it is possible to retransmit more than just a single segment per RTT with TCP SACK. This is accomplished by means of the TCP option shown in Figure 3.8; the actual ACK field in the TCP header remains unchanged. One can immediately see a disadvantage of SACK: it prolongs the header. However, the benefits of the scheme clearly outweigh this issue in most cases.

While a preliminary form of SACK was already specified in RFC 1072 (Jacobson and Braden 1988), the 'valid' specification is RFC 2018 (Mathis et al. 1996). This document allows quite a bit of flexibility regarding SACK usage: it only says that the SACK option may be utilized, recommends using it under all permitted circumstances if it is used at all, and specifically recommends using it for DupACKs. The latter recommendation is easy to explain: in the scenario that is depicted in Figure 3.7, a SACK that is issued in response to segment 4 would explicitly inform the sender that segments 1 and 3 were lost, and segments 2 and 4 were not. Thus, there would be no need to wait for a whole RTT after entering fast retransmit/fast recovery just to be informed that yet another segment must be retransmitted. How exactly the sender behaviour is to be realized is also left open by RFC 2018, but an example procedure is given; in any case, one can expect a significant performance gain from using SACK. This is shown in (Fall and Floyd 1996) with a congestion control extension that reacts in a more-appropriate manner during fast recovery because it has a better notion of the segments that are in flight. This extension is conservative as it strives to enhance the performance on the basis of SACK information but makes only minimal changes to the original Reno behaviour.

Conservative loss recovery

RFC 3517 specifies a SACK-based replacement for the original fast retransmit/fast recovery algorithm in RFC 2581. The mechanism is largely based on the method described in (Fall and Floyd 1996).[16] Roughly, it works as follows:

---

[16]This is not the only SACK-based congestion control enhancement that was proposed in the literature; one notable example of other such work is TCP with *Forward Acknowledgements (TCP FACK)* (Mathis and Mahdavi 1996).

- A data structure (commonly referred to as the *scoreboard*) that stores SACK information from incoming ACKs and a variable called *pipe* are maintained. The latter is an estimate of the current number of segments in flight. In contrast to loss recovery in Reno and NewReno, where *cwnd* is inflated on the basis of an implicitly calculated estimate, the number of segments in flight is explicitly available here. This also means that there needs to be no 'deflation' upon reception of a partial ACK. The *pipe* variable being an essential element of the mechanism, it is often called the *pipe algorithm*.

- Upon arrival of an ACK, any SACKed or cumulatively ACKed bytes are marked in the scoreboard. Since SACK information is advisory, merely SACKed data must not be removed from the retransmission buffer until a corresponding cumulative ACK arrives. This procedure is similar for DupACKs, the first two of which may also trigger the transmission of new segments (i.e. the algorithm allows the using of limited transmit).

- When three DupACKs arrive, a loss-recovery procedure is initiated. As with the fixed NewReno variant that is specified in RFC 3782, this is only done if the sequence number found in these DupACKs does not refer to the same window as in a previous loss-recovery procedure, which is ensured via a variable called *RecoveryPoint*. While this may theoretically seem to be unnecessary with SACK, in practice, it is inevitable: RFC 3517 only specifies a sender update, and it must also work with receivers that do not always send correct SACK information. The loss-recovery procedure basically resembles fast retransmit/fast recovery in RFC 3782, but there are some differences:

  - When the *pipe* variable is updated (i.e. the number of segments in flight is calculated), a packet is assumed to be lost not only when three DupACKs have arrived or the RTO timer expired but also when three segments with a higher sequence number have been SACKed. Also, segments that were retransmitted but are not assumed to be lost are counted twice (we will see that the algorithm has slightly relaxed rules for retransmitting segments).

  - Since there is no *cwnd* inflation/deflation and the number of segments in flight is explicitly available, partial ACKs need no special treatment; just like any additional DupACKs that come in, they cause an update of the scoreboard and the *pipe* variable. Because a general rule says that segments can be sent as long as $(cwnd - pipe)$ is at least an MSS, any incoming segments will allow the sender to transmit a segment.

  - When determining *which* segment to send (i.e. retransmit or choose a new one), the algorithm generally allows retransmitting any unACKed segment which has not been retransmitted before. The idea of this rule is to keep the ACK clock running and thereby decrease the chances for expiry of the RTO timer; it is called a *retransmission last resort* in the specification. Note that this allows the sender to retransmit segments that are not yet assumed to be lost, and an ACK or SACK that covers such a segment is ambiguous: it can indicate that only the original segment has left the network, and it can also indicate that both the original and its duplicate have left the network. This may lead to a wrong pipe estimate; the authors of RFC 3517 consider this to be a rare event with implications that are probably 'limited to corner cases relative to the entire recovery algorithm'.

More-precise calculation of the number of segments in flight and better knowledge about what actually made it to the receiver has some interesting implications. Consider, for example, what happens when a DupACK contains a SACK block that acknowledges reception of 10 segments from a window. In this case, the general rule that segments can be sent whenever (*cwnd* − *pipe*) is greater than an MSS will allow this algorithm to transmit 10 segments right away, which is a good thing not only because TCP SACK then attains higher throughput but also because it keeps the ACK clock running. NewReno, on the other hand, cannot deduce such information from a DupACK and can therefore only inflate *cwnd* by one, which is a good decision (because it is conservative − it is better to underestimate the number of segments in flight than to overestimate it), but is not as efficient as the RFC 3517 loss-recovery algorithm.

Despite all its advantages, this loss-recovery algorithm is not entirely without problems: since it can keep the ACK clock running for quite a long loss event and the RTO timer is only turned off when all outstanding data are (cumulatively) ACKed according to RFC 2988, it is possible for the timer to expire when the sender is still in its loss-recovery phase. RFC 3517 therefore allows re-arming the RTO timer on each retransmission that is sent during recovery − but this has the potential disadvantage of being a more-conservative timer.

Long fat pipes

At this point, it may be worth mentioning that SACK is sometimes regarded as a technology that was specifically designed to enhance the behaviour of TCP over long fat pipes. Clearly, SACK can yield benefits under a wide variety of circumstances. Since the sender window (across all connections) should ideally be equal to the bandwidth × RTT product, using TCP across a link where this product is large also means having a large window. It should be obvious that the RFC 3517 algorithm makes SACK particularly beneficial under such circumstances − the larger the window, the greater the chance of losing several segments from a single window because of congestion. While such an event can easily be handled by SACK, it causes problems with NewReno, which can only retransmit a single segment per RTT. This problem is also aggravated when the RTT is long − thus, we have a greater chance of experiencing a problem that is particularly bad when it happens if we use TCP NewReno (or, even worse, its predecessors) across a long fat pipe.

It is not surprising that TCP SACK was first described in RFC 1072, which is a bundle of updates for enhancing performance across such links.[17] The other features introduced in this RFC are:

- The Timestamps option (see Section 3.3.2), which is particularly beneficial across long fat pipes because such networks require a more-precise RTT estimate, and enhanced precision is what this option yields. The option was called *Echo* (and *Echo*

---

[17]RFC 1072 is perhaps also the first IETF document to state that a network containing a long fat pipe is an 'LFN' (pronounced 'elephan(t)'). This has been quoted so often that it would be a shame to break with this tradition here.

Reply respectively) in RFC 1072. The 'valid' specification of this feature can be found in RFC 1323 (Jacobson et al. 1992). On a side note, RFC 1323 also describes how to use the Timestamps option for a mechanism called *protect against wrapped sequence numbers (PAWS)* – with this option, the receiver can tell a segment from an old cycle through the sequence number space from a new one.

- A 'Window Scale' option which specifies a factor (number of bits) by which the receive window value is shifted in order to fit into the 16-bit field of the TCP header. This is necessary because the maximum possible window size without this option is not large enough to accommodate the bandwidth × RTT product of some large links. Once again, the 'valid' specification of this feature can be found in RFC 1323.

How to use SACK for signalling reception of duplicate segments (called *D-SACK*) was specified in RFC 2883 (Floyd et al. 2000b). This does not require any changes to the option format, and will simply be ignored by a peer that does not support the new specification. Moreover, while segments carrying an old sequence number may be ACKed with this method, it does not clash with PAWS, which luckily requires the receiver to acknowledge a segment even if it detects a sequence number from a previous cycle through the sequence number space. The advantage of D-SACK is more-precise knowledge about duplicates, which may be caused by replication inside the network or unnecessary retransmissions. Like the original SACK specification, RFC 2883 does not go into details about how this feature is to be used – it merely provides a means that is likely to yield benefits if it is utilized in clever ways.

3.4.9   Explicit Congestion Notification (ECN)

The RED active queue management mechanism has gained reasonable acceptance in the Internet community, and is known to be implemented in some routers. We already discussed active queue management in the previous chapter (Section 2.10.2 on Page 27), and we will take a closer look at RED in Section 3.7. Also, towards the end of the previous chapter, we examined the idea of setting a bit which means 'behave as if the packet was dropped' instead of actually dropping a packet – this is Explicit Congestion Notification (ECN) (see Section 2.12.1 on Page 32). Note that ECN *requires* active queue management to be in place – when a queue overflows, setting a bit is simply not an option. Also, while ECN can reduce loss, reacting to loss is still necessary because there can always be traffic bursts that are too large to fit into the buffer of a router, with or without active queue management. Adding ECN functionality to the Internet was proposed in RFC 2481 (Ramakrishnan and Floyd 1999), which was published in January 1999 and obsoleted by RFC 3168 (Ramakrishnan et al. 2001) in September 2001. For ease of explanation, we will start with a description of ECN as it was originally proposed and then look at the most-important changes that were made.

One major difference between ECN as a theoretical construct and ECN in the Internet is the necessity of a second bit that conveys the information 'yes, I understand ECN' to routers. This is necessary because the Internet can only be upgraded in a gradual manner, and it can of course not be assumed that setting a bit in the IP header will make old end systems act as if a packet had been dropped when they do not even understand what the bit means. Therefore, the ECN rule for routers is to follow the traditional model of dropping packets unless they see a bit that tells them that the source understands ECN. There is a second obvious reason for using such a bit – not all transport protocols will support ECN, even after the Internet has been 'upgraded' (e.g. UDP). Hence, a router should discard rather than mark such packets, when necessary. Since it is unclear whether an application that uses UDP will react to congestion, the general recommendation is not to make ECN usage available in the UDP socket interface.

Introducing ECN to the Internet was a major step; after failed attempts with the ICMP 'Source Quench' message, it is the first feasible TCP/IP congestion control solution that incorporates explicit feedback. In theory, the benefit of ECN is loss reduction in the presence of routers that use active queue management, and there are no disadvantages whatsoever. In practice, because of the 'yes, I understand ECN' bit, ECN faces the same deployment problems as any other new technology that leaves traces in the IP header: if it differs from most other packets, some firewalls will drop it. I use a somewhat outdated Linux kernel;[18] were I to recompile it with 'make xconfig' and select 'Networking options', I could see that 'IP: TCP Explicit Congestion Notification support' is disabled by default. Pressing the 'help' button would yield two sentences that explain ECN, and the following text:

> Note that, on the Internet, there are many broken firewalls which refuse connections from ECN-enabled machines, and it may be a while before these firewalls are fixed. Until then, to access a site behind such a firewall (some of which are major sites, at the time of this writing) you will have to disable this option, either by saying N now or by using the sysctl.
>
> If in doubt, say N.

Things may be different with more recent versions of the kernel, and there is some evidence that several routers now support ECN and that some firewalls were upgraded to let packets carrying ECN information through. On a side note, this deployment problem also concerns any new transport protocol, which would use a new protocol number in the IP header that is unknown to firewalls for some time. The common firewall rule 'reject everything unknown' is at odds with the robustness principle (see Section 3.1.1) that is part of the foundation upon which the Internet is built, and it prevents the technology from improving in a seamless manner.

The two ECN bits are bits 6 and 7 in the 'Type of Service' (TOS) field of the IPv4 header and the 'Differentiated Services' (DS) field of the IPv6 header, respectively. Bit 6 is the *ECN Capable Transport (ECT)* bit, which informs routers that a source understands what an ECN bit means (i.e. it will act as if the packet was dropped when it sees the bit). Bit 7 is called *Congestion Experienced (CE)* – this is what we referred to as the 'ECN bit' until now. Since the congestion notification is not directly sent back to the source (as with the ICMP 'Source Quench' message), the receiver also needs a means to convey the information 'I saw a CE bit, so reduce your rate as if the packet had been dropped' back to the sender. For this purpose, one of the originally reserved bits in the TCP header was redefined in RFC 2481 and called *ECN-Echo (ECE)* (see Section 3.1.1). Now, the following problems can occur:

- The ACK carrying an ECN-Echo flag can be dropped. Then, the next ACK could cumulatively acknowledge the reception of the segment that normally would have

---

[18]It is February 2005, so I am slightly ashamed to admit that it is version 2.4.20-8 – but hey, I am supposed to write this book and not fiddle around with my Linux configuration.

been acknowledged by the preceding ACK, and the congestion event is unnoticed. The receiver is therefore required to set the ECE flag in a series of ACK packets.

- A series of ACKs that carry a set ECE flag (this can also be a 'natural' effect: it could be caused by an incoming series of packets with a set CE flag at the receiver) would theoretically cause a series of *cwnd* reductions at the sender. This can be inappropriate: as we have seen in the previous sections, researchers have gone to great lengths to prevent multiple congestion response invocations that are caused by packets from the same window.

The solution is yet another bit, which is to be set by the sender when it reduces its congestion window. This bit is appropriately called *Congestion Window Reduced (CWR)* and is also part of the TCP header (see Section 3.1.1). A sender is not supposed to react to congestion indications more often than once per window of data, and when it does so, it sets the CWR bit. The receiver is supposed to set the ECE flag in ACK packets upon receiving a packet where the CE flag is set until it receives a packet carrying $CWR = 1$. CWR is reliably transmitted to the receiver because the sender will reduce its window in response to a dropped packet anyway and therefore set the flag again if the segment carrying $CWR = 1$ is retransmitted.

With four bits instead of one, ECN is not as simple as it might have seemed to be at first sight; in fact, there is at least one more serious issue with the mechanism. When a packet is dropped, there is no way for a malicious receiver to undo this action and tell the sender that there was no congestion. With ECN, however, a misbehaving receiver could easily inform a sender that everything was all right by just ignoring the CE bit. This would lead the sender to unfairly increase its congestion window further when it should actually be reduced, and increase throughput for the receiver. While congestion control always requires cooperation on the sender's side, the additional danger from uncooperative receivers must be taken into account, as receivers (who want to achieve the best possible performance) and senders (who want to equally share the bandwidth among all clients) often have conflicting interests (Ely et al. 2001). This was solved with a *nonce* – a random number that would need to be guessed by a malicious user – in RFC 3168 (Ramakrishnan et al. 2001).

Robust ECN signalling

The idea is very simple: the sender generates a random number (the nonce) and the receiver sends it back. As a rule, when a router sets the CE bit, it must also delete the nonce (set it to 0), and a sender believes that the information 'there was no congestion' is correct only if the receiver was able to send the nonce back. This way, the receiver would have to guess the random number in order to lie about the congestion state in the network. Figure 3.9 summarizes the whole ECN signalling process: first, a packet carrying $ECT = 1$, $CE = 0$ and a random nonce is sent. When a router detects congestion and its active queue management scheme decides that something must be done, the packet header is checked. If $ECT = 1$, CE is set instead of dropping the packet, and the nonce is cleared. The third step shows how the receiver reflects the congestion event back to the sender using ECE; in addition to setting ECE (shown in the figure), it copies the nonce to ACKs; in step four, the sender is informed about congestion (the nonce is cleared, so $ECE = 0$ would be detected as a lie) and appropriately reduces its congestion window. Then, it sets $CWR = 1$, which informs the receiver to stop setting $ECE = 1$ until there is a new congestion event to report.

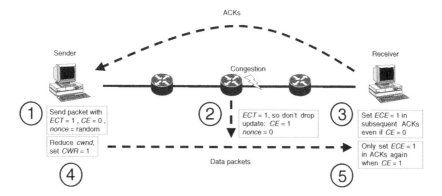

Figure 3.9 How TCP uses ECN

The implementation of the nonce is not as easy as it may seem: since routers should not look at packet headers beyond IP for several reasons (see Section 5.3.1 in Chapter 5 for details), it must obviously be a field in the IP header – but there, bits are costly. Since ECN is only relevant for flows that comprise more than a handful of segments, even a single bit suffices; 32 consecutive packets carrying a single bit each are as valuable as one packet that contains a 32-bit nonce (Ely et al. 2001). That is, while the chance to guess a single random bit is 0.5, the chance to guess n bits in a row is 0.5^n. In RFC 3168, a previously unused bit combination was chosen to be used in such a manner. We have two bits for ECN in the IP header: ECT and CE. At any time, these bits (conjointly called the *ECN field* in RFC 3168) can have the following values:

$ECT = 0$, $CE = 0$: The sender is not ECN-capable and routers will drop packets in case of congestion.

$ECT = 1$, $CE = 0$: The sender is ECN-capable and the packet experienced no congestion so far.

$ECT = 1$, $CE = 1$: The sender is ECN-capable and the packet experienced a congestion event.

Since routers do not update the CE field unless ECT is set, the combination ($ECT = 0$, $CE = 1$) will never occur; this bit combination is therefore a 'vacant spot' that can be used for the nonce. According to the latest definition, the combinations ($ECT = 1$, $CE = 0$) and ($ECT = 0$, $CE = 1$) are called $ECT(0)$ and $ECT(1)$, respectively, and both indicate that the sender is ECN-capable and no congestion occurred. The two values ECT(0) and ECT(1) represent the values 0 and 1 of a single nonce bit, and it will automatically be deleted (destroyed) by a router that sets both ECT and CE to 1 in the presence of congestion. RFC 3168 merely provides the means to work with a nonce; it does not go into details about how this can be realized.

ECN is both beautifully simple and surprisingly complex at the same time – that is, the idea is a simple one, but its implications are numerous, and some of them are quite complicated. We have seen some details about ECN usage here; RFC 3168 elaborates on many further issues including tunnelling, misbehaving routers, attacks with spoofed source addresses and so on. The problem with tunnels, for example, is that ECN support would normally not be possible along a tunnel (i.e. when a packet is encapsulated in another one – why and how to do this for security purposes with IPSec is explained in RFC 2401 (Kent and Atkinson 1998)) unless tunnel endpoints copy the ECN flag to the outer header upon encapsulation and copy it from the outer header to the original packet again at the end of the tunnel. This might raise security issues: should a secure application trust ECN bits in the IP header, marked by untrusted routers inside the core network? Can these bits be used to 'leak' information from the trusted side to the untrusted side or vice versa?

In order to address these concerns, RFC 3168 includes a mechanism for IPSec tunnel endpoints to negotiate whether to use ECN; this way, one can always choose the safe path of simply setting the ECN field to 0 in the outer header. All in all, with its 63 pages, RFC 3168 is a good example of how many difficulties a single-bit idea can bring about, and studying it may be a good exercise when trying to discover all the potential implications of one's own idea.

3.5 Concluding remarks about TCP

All the TCP details described in this chapter are specified in 'standards track' RFCs (shown in Figure 3.10), which essentially means that the IETF recommends their implementation. A perfect TCP implementation would therefore do the following things, all of which influence when a packet is sent into the network (note that this is not a comprehensive list, but just a quick overview of the most important features):

- Set up and tear down a connection whenever a communication flow between two peers starts and ends.

- Limit the sender rate with a window.

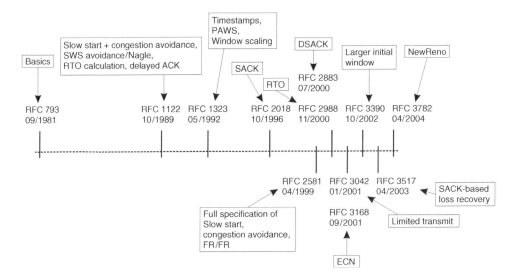

Figure 3.10 Standards track TCP specifications that influence when a packet is sent

- Realize Silly Window Avoidance, in particular the Nagle Algorithm, which adds delay if small amounts of data are Transmitted.

- Delay ACKs at the receiver.

- Go through the slow start, congestion avoidance, fast retransmit and fast recovery algorithms, that is, increase the rate exponentially at the beginning until a threshold is reached, then continue linearly and decrease the rate by half when congestion sets in (only very severe congestion brings a sender back into slow start mode).

- Retransmit dropped packets; such events are detected by expiry of the RTO timer (which is updated using an EWMA process and some variation overhead for every incoming ACK that advances the left edge of the send window, thanks to the Timestamps option), or via three DupACKs. Because of limited transmit, each of these three DupACKs causes the sender to transmit a packet.

- Use the SACK option and the RFC 3517 algorithm, which allows a sender to recover more efficiently from severe loss events, where multiple packets are dropped from a single window.

- Utilize ECN, which has the sender behave as if a packet was dropped when RED only sets a bit to indicate congestion.

All in all, we can see that building a state-of-the art TCP implementation is quite a bit of effort, and not necessarily an easy task. Indeed, there is evidence that TCP implementations frequently malfunction (Paxson et al. 1999); moreover, it is hard to tell how many of the TCP stacks out there support some or all of the described features (and which ones). A related large-scale measurement is described in (Padhye and Floyd 2001), and this effort is ongoing – the accompanying website[19] may be the best place to look for an up-to-date overview of the current state of TCP implementations in the Internet.

We have seen that the standard has changed quite a bit over the years. It is still common among researchers to consider the word 'TCP' as being synonymous with 'TCP Reno', whereas it should actually be synonymous with an implementation that does all the things in the list above. In particular, this becomes a problem when a new mechanism is compared with TCP (e.g. in simulations) – saying that it works better just because it outperforms TCP Reno is unfair, and it is also somewhat impolite towards the many researchers who have devoted quite a bit of their time to update the standard over the years.

This being said, even the most-recent version of TCP is far from being a perfect protocol. If it were, the numerous efforts of researchers who strive to enhance it (some of which we will see in the next chapter) would be pointless. As one can see in Figure 3.10, the protocol has undergone quite a number of changes, and most of them were made in the last couple of years. This diagram must be interpreted with care: for instance, the clustering of changes after the year 1999 only means that quite a number of standards track documents were published that update the protocol. It does not necessarily mean that the changes in these documents were the most important ones – in fact, most researchers will tell you that the most-significant change to TCP was the original addition of congestion control by Van Jacobson, and that was prescribed in 1989. Also, the diagram only shows standards track

---

[19]http://www.icir.org/tbit/

IETF documents; neither seminal papers such as (Jacobson 1988) or (Karn and Partridge 1987) nor experimental RFCs (which may lead to future TCP enhancements and could already be implemented in some stacks) are included.

The documents shown in the figure do not cover all the RFCs that constitute the TCP standard – in (Duke et al. 2004), you will find several others that are not included here because they do not directly influence when TCP sends a packet (and are therefore of minor interest to us). Some RFCs *do* influence when TCP sends a packet, but they are not part of the protocol specification; Path MTU Discovery in RFC 1191 (Mogul and Deering 1990) is one such example.

As mentioned at the beginning of this chapter, the amount of work that has been carried out on problems with TCP in various environments is immense. Many things have been done with the protocol, ranging from formal analyses and simulations to all kinds of real-life tests. While the use of mathematical models was deliberately avoided in this book ((Srikant 2004) is a recommendable source for such things), no TCP description would be complete without the famous *inverse square-root p law*, which is derived from the common sawtooth behaviour of TCP Reno in congestion avoidance mode (neglecting the possibility of RTO timer expiry) under the assumptions that the packet loss ratio p is constant and the receiver acknowledges every packet:

$$T = \frac{s}{RTT}\sqrt{\frac{3}{2p}} \tag{3.6}$$

This equation yields the average sending rate T in bytes per second as a function of the RTT, the packet size s and the packet loss ratio. It tells us that large RTTs and high packet loss ratios lead to poor TCP throughput. The derivation of this model is rather straightforward – see, for instance, (Hassan and Jain 2004) – but it is only a very rough approximation of TCP behaviour. Obviously, the model can be refined, but the further we take this, the more complicated it becomes. A better model that takes slow start into account and is also very well known is the so-called *Padhye equation* (Padhye et al. 1998).

$$T = \frac{s}{RTT\sqrt{\frac{2p}{3}} + t_{RTO}\left(3\sqrt{\frac{3p}{8}}\right)p(1 + 32p^2)} \tag{3.7}$$

which describes the sending rate T in bytes per second as a function of the packet size s, the RTT, the steady-state loss event rate p and the TCP retransmit timeout value t_{RTO} (which is a function of the RTT – we already discussed its calculation in Section 3.3.3).

A system is characterized not only by its features but also by its faults; let us therefore conclude this part of the chapter with some fun. In (Savage et al. 1999), the three following possibilities for a malicious receiver to make the sender increase its rate beyond the 'allowed' maximum are listed:

ACK division: Equation 3.4, which has the sender increase *cwnd* in the congestion avoidance phase, is executed for every ACK and not for every ACKed MSS. Therefore, a misbehaving receiver can make the sender open the window n times faster by splitting an ACK into n smaller ACKs. This attack is also known by the name 'ACK splitting'.

DupACK spoofing: An ACK or DupACK is taken to indicate that a segment has left the network – therefore, *cwnd* is inflated by one segment for each incoming DupACK once the sender is in fast retransmit/fast recovery mode (see Section 3.4.5). While this may no longer work with the 'pipe algorithm' in RFC 3517, simply sending a large number of similar ACKs for the last sequence number received will force TCP Reno or NewReno senders to transmit new segments into the network.

Optimistic ACKing: The TCP congestion control algorithms implicitly assume that a sender will send at most one ACK per segment that it receives and never generate ACKs 'on its own', that is, without waiting for ACKs. As illustrated by Equation 3.7, short RTTs yield a high sending rate. Therefore, if a receiver generates ACKs for data that are not yet received (and therefore artificially shortens the feedback loop, which makes the sender calculate an RTT estimate that is too small), it can obtain higher throughput. One possibility to do this is to send ACKs for data that are not yet received in response to a single incoming segment.

The potential for such attacks illustrates that refining the standard is worthwhile. As a matter of fact, some of the TCP enhancements discussed in the next chapter actually deal with these problems – ABC in Section 4.1.1 is one example, as it prevents the ACK division attack mentioned above.

3.6 The Stream Control Transmission Protocol (SCTP)

The *Stream Control Transmission Protocol* (SCTP) is a standards track IETF protocol (specified in RFC 2960 (Stewart et al. 2000)) that is widely accepted as a reasonable performance enhancement at the transport layer. As such, it is now going through the difficult post-standardization phase of achieving large-scale Internet deployment – firewall designers must be convinced to let SCTP packets through, stacks must be updated and so on. SCTP was originally designed to efficiently transfer telephony signalling data across the Internet, but its features make it attractive for other applications too. These are some of the most important aspects that this protocol adds to the functionality of TCP:

Reliable out-of-order data delivery: When segments 1, 3, 4, 5 and 6 reach a TCP receiver, the data contained in segments 3 to 6 will not be delivered to the application until segment 2 arrives. This effect is caused by the requirement to deliver data in order, and is often referred to as *head-of-line blocking delay*. By allowing applications to relax this constraint, SCTP has the potential to deliver data faster.

Preservation of message boundaries: Delivering out-of-order segments can work only if the data blocks can be clearly identified. In other words, embedding such a function in TCP would not be possible because the protocol is byte-stream oriented. Moreover, giving the application the power to control the elementary data units that are transferred can yield more efficient programs. This concept is known by the name *Application Layer Framing (ALF)* (Clark and Tennenhouse 1990).

Support for multiple separate data streams: Mapping multiple logically independent data streams onto a single TCP connection requires some effort from an application and is inefficient. Even when the streams themselves call for in-order data delivery, this is not necessarily the case for segments that belong to different streams, and head-of-line blocking delay can therefore occur (and adversely affect performance).

A common solution is to utilize multiple TCP connections, but this also means that connection setup and teardown are carried out several times, and that a congestion control 'engine' is active for each connection.

Multihoming: TCP connections are uniquely identified via two IP addresses and two port numbers. In SCTP, the same is achieved with two *sets of* IP addresses and two port numbers, and a connection is called an *association*. This feature was added for the sake of robustness; it can be used to transparently switch from one IP address to another in case of failure (e.g. without an ftp application noticing it). There are at least two potential failure scenarios that can be dealt with: (i) server A fails, and server B automatically takes over, and (ii) a connection along the path to A goes down, SCTP switches to B until dynamic routing fixes the problem (which can take quite a while) and then switches back. Theoretically, such a feature could also be used for load balancing (splitting traffic from/to multiple IP addresses), and some researchers are now looking at this possibility; however, owing to the many unanswered questions that load balancing brings about, such usage was not the intention of the SCTP designers. To notice failure, some communication has to be maintained even if the application has nothing to send (similarly, you cannot know if a server is active without even *ping*ing it) – this is achieved via so-called *heartbeat* messages.

Partial reliability: RFC 3758 (Stewart et al. 2004) describes a 'partial reliability' extension of SCTP, which enables applications to define how persistent the protocol should be in attempting to deliver a message (including 'totally' unreliable data transfer). This allows for multiplexing of unreliable and reliable data streams across the same connection.

It is also worth noting that the protocol has a somewhat unusual packet format. Every SCTP packet begins with a 'common header', which is followed by one or more 'chunks', each of which has its own specific header. Data transmission is therefore achieved by including a 'DATA chunk', while some protocol functions (e.g. SACK) can be activated by embedding the corresponding control chunks.

Congestion control in SCTP is largely similar to TCP; in particular, while some of the features listed above (e.g. reliable out-of-order data delivery) do not influence congestion control at all, some of them do. Here are some noteworthy details regarding SCTP congestion control:

- SACK usage is mandated.

- The ACK division attack that was explained in the previous section is prevented by appropriately incorporating the *size* of an acknowledged DATA chunk in the *cwnd* update procedure.

- A distinct set of congestion control parameters is maintained for each destination IP address that is part of an association. The parameters should decay if an IP address is not used for a long time. In any case, whenever the multihoming feature described above switches to an IP address that has not been used before, slow start is performed.

- From a congestion control perspective, sending multiple independent data streams over one SCTP association is as aggressive as one TCP sender, and the common

TCP equivalent – using n separate connections – is as aggressive as n TCP senders. Essentially, the congestion control advantage from multiplexing flows in such a manner is similar to sharing TCP state with the 'Congestion Manager' (see Section 4.2.2 in the next chapter).

- Sending a single totally unreliable data stream with SCTP resembles using the 'Datagram Congestion Control Protocol' with CCID 2, 'TCP-like congestion control' (see Section 4.5.2 in the next chapter).

This concludes our discussion of SCTP; for further details about this protocol, consult RFC 2960 or (Stewart and Xie 2001).

3.7 Random Early Detection (RED)

RFC 2309 (Braden et al. 1998) says:

> Internet routers should implement some active queue management mechanism to manage queue lengths, reduce end-to-end latency, reduce packet dropping, and avoid lock-out phenomena within the Internet.

> The default mechanism for managing queue lengths to meet these goals in FIFO queues is Random Early Detection (RED) (Floyd and Jacobson 1993). Unless a developer has reasons to provide another equivalent mechanism, we recommend that RED be used.

In the first statement above, 'lock-out phenomena' refers to what we called *phase effects* in Section 2.9. As already mentioned in the previous chapter, active queue management schemes make packet dropping or 'marking' (e.g. by setting $CE = 1$ in the IP header) decisions based on the queue behaviour, with the goal of avoiding such phase effects as well as keeping the queue size low while allowing occasional bursts. You may also recall that such functionality typically requires some randomness – hence the name of the default mechanism that RFC 2309 recommends for Internet routers. While it is extremely hard to tell which mechanisms are now really in use in the Internet (after all, the only effects seen by end systems are packet drops and ECN marks) and there are a significant number of proposals for alternatives (see Section 4.4 in the next chapter), RED is probably still the most widely deployed one.

Every time a new packet arrives at the router[20], this algorithm calculates the average queue length using an EWMA process (the same method that is used for RTO calculation in TCP end nodes):

$$avg = (1 - w_q)avg + w_q q \tag{3.8}$$

where avg is the average queue length estimate, q is the instantaneous queue length and w_q is a weighting factor that controls how fast the moving average adapts to fluctuations. Then, it is compared to two thresholds called min_{th} and max_{th}. If the average queue size is less than the minimum threshold min_{th}, nothing happens. If it is greater than the maximum

---

[20]Since this mechanism is controlled by long-term average behaviour, it will also work if this interval is longer (Jacobson 1998).

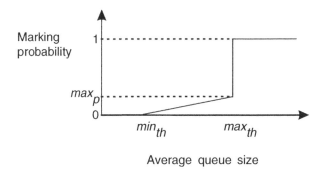

Figure 3.11 The marking function of RED

threshold max_{th}, every arriving packet is marked.[21] This ensures that the average queue size does not significantly exceed the maximum threshold.

Randomness comes into play only when the average queue length is *between* min_{th} and max_{th} – then, the probability of dropping a packet will be between zero and the maximum marking probability max_p, and it will directly be proportional to the average queue length. In other words, when the average queue length grows beyond min_{th}, the marking probability rises linearly from zero to max_p, which is when the average queue length will grow beyond max_{th} and all packets will be marked. This is shown in Figure 3.11.

The reason for choosing a probability function that linearly grows with the number of packets that arrive is that this ensures relatively equal spacing of packet marking events through time; this is explained in detail in (Floyd and Jacobson 1993). It was a goal to avoid clustering of these events – for example, marking a lot of packets at the beginning of a congestion phase and hardly marking any towards its end is not desired. In order to understand why such clustering is problematic, consider a simple FIFO queue that drops all incoming packets whenever its length is exceeded. This is an extreme case of clustering 'marking' (in this case, dropping) events, and it leads to phase effects. If we want to avoid them, we therefore cannot allow such clustering to occur.

RED has a problem that is well known – its parameters are quite hard to tune. Let us examine their influence and how they should ideally be set:

w_q: This determines the reactiveness of the EWMA process to traffic fluctuations. In order to figure out how to tune this parameter, it is crucial to understand why the average (and not the instantaneous) queue length is used in RED: the goal is to filter out short sporadic bursts and only react to persistent congestion. Now consider what would happen if w_q were 1: only the instantaneous queue would be used, and the impact of preceding values would be completely eliminated. Setting this parameter to 0, on the other hand, would mean that the average queue length would remain fixed at some old value and not react to queue fluctuations at all. In (Floyd and Jacobson 1993), a lower and an upper bound for w_q that depends on the size of bursts that one wants to allow are derived.

---

[21] or dropped – note that the original RED algorithm description in (Floyd and Jacobson 1993) leaves the specific 'marking' action open; for the sake of simplicity, we will only talk about 'marking' packets from now on. It is assumed that most flows will reduce their rates in response to this 'marking' action.

min_{th}, max_{th}: These values depend on the desired average queue size. As already mentioned, marking any packets that arrive when the average queue length exceeds max_{th} prevents the queue from growing far beyond this upper limit. In other words, setting this parameter to a small value will lead to a small queue (and thus short delay). On the other hand, the parameter min_{th} depends on the burstiness of traffic – if fairly bursty traffic should be accommodated, it must be set to a rather large value – and at the same time, ($max_{th} - min_{th}$) should not be too small to allow for the randomness to take effect. If, for example, the two parameters were equal, there would be no randomness whatsoever in the scheme, and traffic phase effects would not be countered.

max_p: This parameter controls how likely it is for a packet to be discarded when the average queue length is between min_{th} and max_{th}; it should be small because the general goal of RED is not to drop a large number of packets as soon as min_{th} is exceeded but only drop a packet every now and then, thereby forcing senders to reduce their rates. It is recommended to set it to 0.1.

RED parameter settings have been extensively discussed; Sally Floyd provides some hints that are more up to date than (Floyd and Jacobson 1993) on a related web page[22], where she also makes it clear that it is very hard to find *optimal* values for these parameters as they would have to depend on the typical RTTs in the system. We will see in Section 4.4 that alternatives to RED that eliminate these problems have been proposed in the literature. In any case, the mechanism can be expected to be beneficial even with suboptimal parameter choices; this is particularly true if RED is used with the 'gentle' mode that is also suggested by Sally Floyd on her web page. The underlying idea is to avoid the sudden 'jump' to 1 as the average queue size approaches max_{th}, and to have RED slowly increase the packet drop rate as it approaches $2 * max_{th}$ instead, as shown in Figure 3.12. The advantages of RED are manifold:

- It eliminates traffic phase effects.

- It allows occasional bursts but still keeps the average queue size low, which means that sources experience less delay, making them more reactive to congestion. Also, they will react earlier, which in turn keeps the queue size low.

- Since the probability that a packet from a sender is dropped (or that the sender is notified via ECN) is roughly proportional to the rate of the sender, RED 'punishes' unresponsive senders that send at a high rate to some degree.

- Since RED usually drops a single packet from a flow during the onset of congestion, whereas a normal FIFO queue without RED usually drops a series of packets per flow (again a phase effect), chances are better that a TCP sender can recover from a RED congestion event. This is because a single missing packet will typically trigger fast retransmit/fast recovery whereas a large series of missing packets may prevent the necessary number of DupACKs from arriving, thereby increasing the chance of a timeout.

---

[22] http://www.icir.org/floyd/red.html#parameters

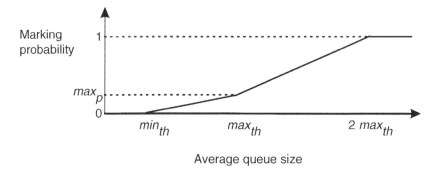

Figure 3.12 The marking function of RED in 'gentle' mode

- ECN, which has the advantage of causing less (and in some cases *no*) loss, can only work with an active queue management scheme such as RED.

Sally Floyd maintains some information regarding implementation experiences with RED on her web page. Given the facts on this page and the significant number of well-known advantages, there is reason to hope that RED (or some other form of active queue management) is already widely deployed, and that its use is growing. In any case, there is no other IETF recommendation for active queue management up to now – so, if your packets are randomly dropped or marked, chances are that it was done by RED or one of its variants.

3.8   The ATM 'Available Bit Rate' service

ATM was an attempt to build a new network that supports multimedia applications such as pay-per-view or video conferencing through differentiated and accordingly priced service classes. It is a highly complex technology that was defined with its own three-dimensional layer model, and it was supposed to provide services at all layers of the stack. Underneath it all, *cells* – link layer data units with a fixed size of 53 bytes, five of which constitute the header – are sent across fibre links. These cells are used to realize circuit-like behaviour via time division multiplexing. If, for example, every fifth cell along a particular set of links is devoted to a particular source/destination pair, the provisioned data rate can be precisely calculated; this results in a strictly connection-oriented service where the connection behaves like a leased line. Cells must be small in order to enable provisioning of such services with a fine granularity. Specifically, the services of ATM are as follows:

Constant Bit Rate (CBR) for real-time applications that require tightly constrained delay variation.

Real-Time Variable Bit Rate (rt-VBR) for real-time applications that require tightly constrained delay variation and transmit with a varying data rate.

Non-Real-Time Variable Bit Rate (nrt-VBR) for applications that have no tight delay or delay variation constraints, may want to send bursty traffic but require low loss.

Unspecified Bit Rate (UBR) for applications such as email and file transfer (this is the ATM equivalent of the Internet 'best effort' service).

Guaranteed Frame Rate (GFR) for applications that may require a minimum rate (but not delay) guarantee and can benefit from accessing additional bandwidth dynamically available in the network.

Available Bit Rate (ABR) which is a highly sophisticated congestion control framework. We will explain it in more detail below.

Today, the once popular catch phrase 'ATM to the desktop' only remains a reminiscence of the better days of this technology. In particular, the idea of bringing ATM services to the end user never really made it in practice. There are various reasons for this; one fundamental problem that might have been the primary reason for ATM QoS to fail is the fact that differentiating between end-to-end flows in all involved network nodes does not scale well. Nowadays, ATM is still used in some places, but almost only as a link layer technology for transferring IP packets over fibre links in conjunction with the UBR or ABR service. In the Internet of today, we can therefore encounter ATM ABR as some kind of link layer congestion control functionality that runs underneath IP.

First and foremost, the very fact that ATM ABR is a *service* is noteworthy: congestion control can indeed realize (or be regarded as) a service. Specifically, ABR is a cheap service that just gives a source the bandwidth that is not used by any other services (hence the name); it is not intended to support real-time applications. As users of other services increase their load, ABR traffic is supposed to 'give way'. One additional advantage for applications using this service is that by following the 'rules' they greatly decrease their chance of experiencing loss. The underlying element of this service is the concept of *Resource Management (RM) cells*. These are the most interesting fields they carry:

BECN Cell (BN): This flag indicates whether the cell is a Backward ECN cell or not. BECN cells – a form of choke packets (see Section 2.12.2) – are generated by a switch,[23] whereas non-BECN RM cells are generated by senders (and sent back by destinations).

Congestion Indication (CI): This is an ECN bit (see Section 2.12.1).

No Increase (NI): This flag informs the sender whether it may increase its rate or not.

Explicit Rate (ER): This is a 16-bit number that is used for explicit rate feedback (see Section 2.12.2).

This means that ATM ABR provides support for a diversity of explicit feedback schemes at the same time: ECN, choke packets and explicit rate ER feedback. All of this is specified in (ATM Forum 1999), where algorithms for sources, destinations and switches are also outlined in detail. This includes answers to questions such as when to generate an RM cell, how to handle the NI flag, and how to specify a *minimum cell rate* (there is also a corresponding field for this in RM cells). Many of these issues are of minor interest; the part

---

[23] You can think of an ATM switch as a router; these devices are called *switches* to underline the fact that they provide what 'looks and feels' like a leased line to end systems.

that received the greatest attention is, without doubt, handling of the ER field. Basically, ATM ABR ER feedback works as follows:

- The source sends RM cells to the destination at well-defined time intervals; the ER field of these cells carries a requested rate (smaller or equal to the initially negotiated 'Peak Cell Rate' (PCR)).

- Upon reception of the RM cell, each switch calculates the maximum rate that it wants to allow a source to use. If its calculated rate is smaller than the value that is already in the field, then the ER field of the RM cell is updated.

- The destination reflects the RM cell back to the sender.

- The sender always maintains a rate that is smaller or equal to the value in the most recently received ER field.

Notably, intermediate nodes can themselves work as source or destination nodes (they are then called *Virtual Source* and *Virtual Destination*). This effectively divides an ABR connection into a number of separately controlled segments and turns ABR into some sort of a hop-by-hop congestion control scheme. Thus, ATM ABR supports all the explicit feedback schemes that were presented in Section 2.12 of Chapter 2.

3.8.1   Explicit rate calculation

The most-interesting part that remains to be explained is the switch behaviour. While there is no explicit rule that specifies what fairness measure to apply, the recommended default behaviour for the case when sources do not specify a minimum cell rate is to use max–min fairness (see Section 2.17.1). Since the specification is open enough to allow for a large diversity of ER calculation methods provided that they attain (at least) a max–min fair rate allocation, a newly developed mechanism that works better than an already existing one can theoretically be used in an ATM switch right away without violating the standard. Since creating such a mechanism is not exactly an easy task, this led to an immense number of research efforts. Since the ATM ABR specification document (ATM Forum 1999) was updated a couple of times over the years before it reached its final form, it also contains an appendix with a number of example mechanisms. These are therefore clearly the most-important ones; let us now take a closer look at the problem and then examine some of them.

It should be straightforward that one can theoretically do better than a mechanism like TCP if there is more explicit congestion information available to end nodes. The main problem with such schemes is that they typically require switches to carry out quite sophisticated calculations in order to achieve max–min fairness. This is easy to explain: as we already mentioned in Section 2.17.1, in the simple case of only one switch, dividing the bandwidth according to this fairness measure means that n flows would each be given exactly b/n, where b is the available bandwidth. In order to calculate b/n, a switch must typically know (or be able to estimate) n – and this is where the problems begin. Actually counting the flows would require remembering source–destination pairs, which is per-flow state; however, we have already identified per-flow state as a major scalability hazard, in Section 2.11.2, and this is perhaps the biggest issue with ATM ABR. ATM, in general, has been said not to scale well, and it is clearly not a popular technology in the IETF.

One scheme that explicitly requires calculating the number of flows in the system is *Explicit Rate Indication for Congestion Avoidance*(ERICA), which is an extension of an original congestion avoidance mechanism called *OSU scheme* (OSU stands for 'Ohio State University'). It first calculates the input rate to a switch as the number of received cells divided by the length of a measurement interval. Then, a 'load factor' is calculated by dividing the input rate by a certain *target rate* – a value that is close to the link capacity, but leaves a bit of overhead (e.g. 95%). There are several variants of this mechanism (one is called 'ERICA+'), but according to (ATM Forum 1999), in its simplest form, a value called *Vcshare* is calculated by dividing the current cell rate of the flow (another field in RM cells) by the load factor, and a 'fair share' (the minimum rate that a flow should achieve) is calculated by dividing the target rate by the number of flows. Then, the ER field in the RM cell is set to the maximum of these two values. Note that fair share calculation requires knowledge of the number of flows – and therefore per-flow state. In other words, in the form presented here, ERICA cannot be expected to scale too well.

Congestion Avoidance using Proportional Control (CAPC) calculates a load factor just like ERICA. Determining the ERs is done by distinguishing between *underload state*, where the load factor is smaller than one, that is, the target rate is not yet reached, and *overload state*, where the load factor is greater than one. In the first case, the fair share is calculated as

$$fair\ share = fair\ share * \min(ERU, 1 + (1 - load\ factor) * R_{up}) \qquad (3.9)$$

whereas in the second case, the fair share is calculated as

$$fair\ share = fair\ share * \max(ERF, 1 + (load\ factor - 1) * R_{dn}) \qquad (3.10)$$

where R_{up} and R_{dn} are 'slope parameters' that determine the speed (reactiveness) of the control and *ERU* and *ERF* are used as an upper and lower limit, respectively. R_{up} and R_{dn} represent a trade-off between the time it takes for sources to saturate the available bandwidth and the robustness of the system against factors such as load fluctuations and the magnitude of RTTs.

CAPC achieves convergence to efficiency by increasing the rate proportional to the amount by which the traffic is less than the target rate and vice versa. The additional scaling factors ensure that fluctuations diminish with each update step while the limits keep possible outliers within a certain range. This idea is shown in Figure 3.13, which depicts the function

$$f(x) = \begin{cases} x + R_{up}(target - x) & \text{if } x < target \\ x - R_{dn}(x - target) & \text{if } x > target \end{cases} \qquad (3.11)$$

with $R_{dn} = 0.7$, *target* $= 7$ and different values for R_{up}: as long as the scaling factors R_{up} and R_{dn} are tuned in a way that prevents $f(x)$ from oscillating, the function converges to the target value. This is a simplification of CAPC, but it suffices to see how proportional adaptation works.

Another noteworthy mechanism is the *Enhanced Proportional Rate Control Algorithm (EPRCA)*, which uses an EWMA process to calculate a 'Mean Allowed Cell Rate' (MACR):

$$MACR = (1 - \alpha)MACR + \alpha CCR \qquad (3.12)$$

where *CCR* is the current cell rate found in the RM cell and α is generally chosen to be 1/16, which means that it weights the MACR 15 times more than the current cell rate. The

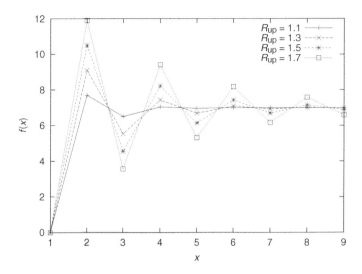

Figure 3.13 Proportional rate adaptation as in CAPC. Reproduced by kind permission of Springer Science and Business Media

fair share – which is not to be exceeded by the value of the ER field in the RM cell – is calculated by multiplying MACR with a 'Down Pressure Factor' which is smaller than 1 and recommended to be 7/8 in (ATM Forum 1999). This scheme, which additionally monitors the queue size to detect whether the switch is congested and should therefore update the ER field or not, was shown not to converge to fairness under all circumstances (Sisalem and Schulzrinne 1996).

Researchers have taken ATM ABR rate calculation to the extreme; mechanisms in the literature range from ideas where the number of flows is estimated by counting RM cells (Su et al. 2000) to fuzzy controllers (Su-Hsien and Andrew 1998). Coming up with such things makes sense because the framework is open enough to support any kind of complex methods as long as they adhere to the rule of providing some kind of fairness. This did not render the technology more scalable or further its acceptance in the IETF; the idea of providing an ABR service to an end user was given up a long time ago. Nowadays, ATM is used to transfer IP packets just because it is a fibre technology that is already available in some places. There are, however, some pitfalls when running IP and especially TCP over ATM.

3.8.2   TCP over ATM

One problem with TCP over ATM is that the fundamental data unit is much smaller than a typical IP packet, and this data unit is acted upon. That is, if an IP packet consists of 100 ATM cells and only one of them is dropped, the complete IP packet becomes useless. Transmitting the remaining 99 cells is therefore in vain, and it makes sense to drop all remaining cells that belong to the same IP packet as soon as a cell is dropped. This mechanism is called *Partial Packet Discard (PPD)*. In addition to requiring the switch to maintain per-flow state, this scheme has another significant disadvantage: if the cell that was dropped is, say, cell number 785, this means that 784 cells were already uselessly transferred (or enqueued) by the time the switch decides to drop this cell.

A well-known solution to this problem is to realize *Early Packet Discard (EPD)* (Romanow and Floyd 1994). Here, a switch decides to drop all cells that belong to a packet when a certain degree of congestion is reached (e.g. a queue threshold is exceeded). Note that this mechanism, which also requires the switch to maintain per-flow state, constitutes a severe layer violation – but this is in line with newer design principles such as ALF (Clark and Tennenhouse 1990).

Congestion control implications of running TCP over ABR are a little more intricate. When TCP is used on top of ABR, a control loop is placed on top of another control loop. Adverse interactions between the loops seem to be inevitable; for instance, the specification (ATM Forum 1999) leaves it open for switches to implement a so-called use-it-or-lose-it policy, where sources that do not use the rate that they are allowed to use at any time may experience significantly degraded throughput. TCP, which uses slow start and congestion avoidance to probe for the available bandwidth, is a typical example of one such source – it hardly ever uses all it could. This may also heavily depend on the switch mechanism that is in place; simulations with ERICA indicate that TCP performance is not significantly degraded if buffers are large enough (Kalyanaraman et al. 1996). On the other hand, it seems that TCP can work just as well over UBR, and that the additional effort of ABR does not pay off (Ott and Aggarwal 1997).

4

Experimental enhancements

This chapter is for researchers who would like to know more about the state of the art as well as for any other readers who are interested in developments that are not yet considered technically mature. The scope of such work is immense; you will, for instance, hardly find a general academic conference on computer networks that does not feature a paper about congestion control. In fact, even searching for general networking conferences or journal issues that do not feature the word 'TCP' may be quite a difficult task. Congestion control research continues as I write this – this chapter can therefore only cover some select mechanisms. The choice was made using three principles:

1. *Mechanisms that are likely to become widely deployed within a reasonable timeframe should be included.* It seems to have become a common practice in the IETF to first publish a new proposal as an experimental RFC. Then, after some years, when there is a bit of experience with the mechanism (which typically leads to refinements of the scheme), a follow-up RFC is published as a standards track RFC document. While no RFC status can guarantee success in terms of deployment, it is probably safe to say that standards track documents have quite good chances to become widely used. Thus, experimental IETF congestion control work was included.

2. *Mechanisms that are particularly well known should be included as representatives for a certain approach.*

3. *Predominantly theoretical works should not be included.* This concerns the many research efforts on mathematical modelling and global optimization, fairness, congestion pricing and so on. If they were to be included, this book would have become an endless endeavour, and it would be way too heavy for you to carry around. These are topics that are broad enough to fill books of their own – as mentioned before, examples of such books are (Courcoubetis and Weber 2003) and (Srikant 2004).

We have already discussed some general-purpose TCP aspects that could be considered as fixes for special links (typically LFPs) in the previous chapter; for example, SACK is frequently regarded as such a technology. Then again, in his original email that introduced fast retransmit/fast recovery, Van Jacobson also described these algorithms as a fix

Network Congestion Control: Managing Internet Traffic   Michael Welzl
© 2005 John Wiley & Sons, Ltd

for LFPs – which is indeed a special environment where they appear to be particularly beneficial. It turns out that the same could be said about many mechanisms (stand-alone congestion control schemes and small TCP tweaks alike) even though they are generally applicable and their performance enhancements are not limited to only such scenarios. For this reason, it was decided not to classify mechanisms on the basis of the different network environments, but to group them according to the functions instead. If something works particularly well across, say, a wireless network or an LFP, this is mentioned; additionally, Table 4.3 provides an applicability overview.

The research efforts described in this chapter roughly strive to fulfil the following goals, and this is how they were categorized:

- Ensure that TCP works the way it should (which typically means making it more robust against all kinds of adverse network effects).

- Increase the performance of TCP without changing the standard.

- Carry out better active queue management than RED.

- Realize congestion control that is fair towards TCP (TCP-friendly) but more appropriate for real-time multimedia applications.

- Realize congestion control that is more efficient than standard TCP (especially over LFPs) using implicit or explicit feedback.

Since the first point in this list is also the category that is most promising in terms of IETF acceptance and deployment chances, this is the one we start with.

4.1   Ensuring appropriate TCP behaviour

This section is about TCP enhancements that could be regarded as 'fixes' – that is, the originally intended behaviour (such as ACK clocking, halving the window when congestion occurred and going back to slow start when the 'pipe' has emptied) remains largely unaltered, and these mechanisms help to ensure that TCP really behaves as it should under all circumstances. This includes considerations for malicious receivers as well as solutions to problems that became more important as TCP/IP technology was used across a greater variety of link technologies. For example, one of these updates fixes the fact that the standard TCP algorithms are a little too aggressive when the link capacity is high; also, there is a whole class of detection mechanisms for the so-called spurious timeouts – timeouts that occur because the RTO timer expired as a result of sudden delay spikes, as caused by some wireless links in the presence of corruption. Generally, most of the updates in this section are concerned with making TCP more robust against such environment conditions that might have been rather unusual when the original congestion control mechanisms in the protocol were contrived.

4.1.1   Appropriate byte counting

As explained in Section 3.4.4, the sender should increase its rate by one segment per RTT in congestion-avoidance mode. It was also already mentioned that the method of increasing *cwnd* by *MSS* ∗ *MSS* /*cwnd* whenever an ACK comes in is flawed. For one, even if the receiver immediately ACKs arriving segments, the equation increases *cwnd* by slightly less than a segment per RTT. If the receiver delays its ACKs, there will only be half as many of them – which means that this rule will then make the sender increase its rate by at most one segment every two RTTs. Moreover, as we have seen in Section 3.5, a sender can even be tricked into increasing its rate much faster than it should by sending, say, 1000 one-byte-ACKs instead of acknowledging 1000 bytes at once.

The underlying problem of all these issues is the fact that TCP does not increase its rate on the basis of the number of bytes that reach the receiver but it does so on the basis of the number of ACKs that arrive. This is fixed in RFC 3465 (Allman 2003), which describes a mechanism called *appropriate byte counting (ABC)*, and this is exactly what it does: counts bytes, not ACKs. Specifically, the document suggests to store the number of bytes that have been ACKed in a 'bytes_acked' variable, which is decremented by the value of *cwnd*. Whenever it is greater than or equal to the value of *cwnd*, *cwnd* is incremented by one MSS. This will open *cwnd* by at most one segment per RTT and is therefore in conformance with the original congestion control specification in RFC 2581 (Allman et al. 1999b).

Slow start is a slightly different story. Here, *cwnd* is increased by one MSS for every incoming ACK, but again, receivers that delay ACKs experience different performance than receivers that send them right away, and it would seem more appropriate to increase *cwnd* by the number of bytes acked (i.e. two segments) in response to such ACKs. However, simply applying byte counting here has the danger of causing a sudden burst of data, for example, when a consecutive series of ACKs are dropped and the next ACK cumulatively acknowledges a large amount of data. RFC 3465 therefore suggests imposing an upper limit *L* on the value by which *cwnd* could be increased during slow start. If *L* equals one MSS, ABC is no more aggressive than the traditional rate update mechanisms but it is still more appropriate for some reasons.

One of them is that ABC with *L* = *MSS* still manages to counter the aforementioned ACK splitting attack. The fact that it is potentially *more* conservative than the traditional rate-update scheme if very few data are transferred is another reason. Consider, for example, a Telnet connection where the Nagle algorithm is disabled. What happens in such a scenario is that the slow-start procedure is carried out as usual (one segment is sent, one ACK is returned, two segments are sent, two ACKs are returned, and so on), but the segments are all very small, and so is the amount of data acknowledged. This way, *cwnd* can reach quite a high value because it does not necessarily reflect the actual network capacity without ABC. If the user now enters a command that causes a large amount of data to be transferred, this will cause a sudden undesirable data burst.

One could also use a greater value for *L* – but the greater its value, the smaller the impact of this limit. Recall that it was introduced to avoid sudden bursts of traffic from a series of lost ACKs. One choice worth considering is to set *L* to 2 ∗ *MSS*, as this would mitigate the impact of delayed ACKs – by allowing a delayed ACK to increase *cwnd* just like two ACKs would, this emulates the behaviour of a TCP connection where the receiver immediately acknowledges all incoming segments. The disadvantage of this method is that it slightly increases what is called *micro burstiness* in RFC 3465: in response to a single delayed ACK, the sender may now increase the number of segments that it transmits by two segments. Also, it has the sender open *cwnd* by a greater value per RTT. This somewhat less cautious method of probing the available bandwidth slightly increases the loss rate experienced with ABC-enabled senders that use $L = 2 * MSS$, which makes this choice somewhat critical.

Finally, L should always be set to one MSS after a timeout, as it is common that a number of out-of-order segments that were buffered at the receiver are suddenly ACKed in such a situation. However, these ACKs do not indicate that such a large amount of data has really left the 'pipe' at this time.

4.1.2 Limited slow start

One potential problem of TCP has always been its start-up phase: it rather aggressively increases its rate up to a *ssthresh* limit, which does not relate to the congestion state of the network. There are several proposals to change this initial behaviour – for example, in addition to the fast retransmit update that is now known as 'NewReno', a method for finding a better initial *ssthresh* value was proposed in (Hoe 1996). The underlying idea of this was to assume that the spacing of initial ACKs would indicate the bottleneck link capacity (see Section 4.6.3); in (Allman and Paxson 1999), such schemes were shown to perform poorly unless complemented with additional receiver-side mechanisms. According to this reference, it is questionable whether estimating the available bandwidth at such an early connection stage is worth the effort, given the complexity of such an endeavour.

While it is unclear whether dynamically calculating *ssthresh* at start-up is a good idea, it seems to be obvious that a sender that has an extremely large window (say, thousands of segments) should not be allowed to blindly double its rate. In the worst case, a sender in slow start can transmit packets at almost twice the rate that the bottleneck link can support before terminating. If the window is very large just before slow start exceeds the bottleneck, this could not only overwhelm the network with a flood of packets but also cause thousands of packets to be dropped in series. This, in turn, could cause a timeout and bring the sender back into slow-start mode again.

For such cases, RFC 3742 (Floyd 2004) describes a simpler yet beneficial change to slow start: an initial parameter called max_*ssthresh* is introduced. As long as *cwnd* is smaller or equal to max_*ssthresh*, everything proceeds normally, but otherwise, *cwnd* is increased in a more-conservative manner – this is called *limited slow start*. The exact *cwnd* update procedure for cases where *cwnd* exceeds max_*ssthresh* is:

$$K = int(cwnd/(0.5 * \text{max\_ssthresh})) \tag{4.1}$$

$$cwnd = cwnd + int(MSS/K) \tag{4.2}$$

RFC 3742 recommends setting max_*ssthresh* to 100 MSS. Let us consider what happens if *cwnd* is 64 MSS (as a result of updating an initial two MSS – sized window five times): 64 segments are sent, and *cwnd* is increased by one for each of the ACKs that these segments cause. At some point, *cwnd* will equal 101 MSS and therefore exceed max_*ssthresh*. Then, K will be calculated; *cwnd*/50 yields 2.02, which will be cut down to 2 by the *int* function. Thus, *cwnd* will be increased by MSS/2, until K is at least 3. From then on, *cwnd* will be increased by MSS/3 and so on. The greater the *cwnd*, the smaller the increase factor becomes; every RTT, *cwnd* increases by approximately $MSS * \text{max\_ssthresh}/2$. This limits the transient queue length from slow start.

Experiments with the 'ns' network simulator (see Appendix A.2) have shown that limited slow start can reduce the number of drops and thereby improve the general performance of TCP connections with large RTTs. Similar experiences were made with real-life tests using the Linux 2.4.16 Web100 kernel.[1]

4.1.3 Congestion window validation

Normally, when considering the congestion control mechanisms of TCP, it is assumed that a sender is 'greedy', which means that it always sends as much as it can. The rules specify that a sender cannot send *more* than what the window allows, but it is generally acceptable to send less. When an application has nothing to send for more than an RTO, RFC 2581 and (Jacobson 1988) suggest that the TCP sender should go back to slow start. Since this is not a mandatory rule, not all implementations do this. If an implementation does not follow this, it can suddenly generate a large burst of packets after a long pause, which may significantly contribute to congestion and cause several packets to be lost because its behaviour has nothing to do with the congestion state of the network. This problem can also occur with 'greedy' senders such as file transfers, for example,. when several files are downloaded across a single TCP connection and the receiving application asks the user where to store the data whenever a file arrives.

Going back to slow start as proposed in RFC 2581 resembles the 'use-it-or-lose-it' policy of ATM switches that we already discussed in Section 3.8.2. On the one hand, such behaviour is appropriate because sending nothing for more than an RTO means that the sender assumes that the 'pipe' has emptied; on the other hand, the fact that the application decided not to transmit any data does not say anything about the state of congestion in the network. In the case of severely limited applications such as Telnet, which only generates traffic when the user decides to type something, this can lead to quite an inefficient use of the available network capacity.

RFC 2861 (Handley et al. 2000b) proposes to *decay* TCP parameters instead of resetting them in such a radical manner. In particular, the idea is to reduce *cwnd* by half for every RTT that a flow has remained idle, while *ssthresh* is used as a 'memory' of the recent congestion window. In order to achieve this, it is set to the maximum of its current value and 3/4 of the current *cwnd* (that is, in between the current value of *cwnd* and the new one) before halving *cwnd* as a result of idleness. The goal of this procedure is to allow an application to quickly recover most of its previous congestion window after a pause.

It is also possible that an application does not entirely stop sending for an RTT or more but constantly transmits slightly less than what *cwnd* allows. In this case, there is some probing of the network state going on, but not at the desired rate (sampling frequency). That is, a more-conservative decision must be taken than in cases where *cwnd* is always fully utilized. RFC 2861 says that the sender should keep track of the maximum amount of the congestion window used during each RTT, and that the actual value of *cwnd* should decay to midway between its original value and the largest one that was used every RTT. In any case, *cwnd* should not be increased unless the sender fully uses it. There is pseudo-code in RFC 2861 that makes it clear how exactly these things are to be done – this concerns the detection that an RTT has passed, among other things.

---

[1] http://www.web100.org

4.1.4   Robust ECN signalling

The idea of an ECN nonce – a random number from the sender that would need to be guessed by a malicious receiver in order to lie about an ECN mark – was already introduced in Section 2.12.1. We have seen that RFC 3168 provides a sender with a means to realize a one-bit nonce which is automatically erased by routers which set CE = 1 via the two bit combination ECT(0) and ECT(1). It was also mentioned that this RFC does not go into details about usage of the nonce.

This is what the experimental RFC 3540 (Spring et al. 2003) takes care of; it explains how to generate a nonce and how to deal with it on the receiver side. The sender randomly selects either ECT(0) or ECT(1). Additionally, it calculates the sum (as an XOR) of the generated nonces whenever a new segment is sent and maintains a mapping from sequence numbers in segments to the corresponding calculated nonce sum. This is the nonce value that is expected in ACKs that carry the same sequence number. For each ACK, the receiver calculates a one-bit nonce sum (as an exclusive-or) of nonces over the byte range represented by the acknowledgement. This value is stored in a newly defined bit (bit number seven in byte 13 of the TCP header), or the rightmost bit of the 'Reserved' field (just to the left of CWR) in Figure 3.1.

The reason for using a sum is to prevent a receiver from hiding an ECN mark (and therefore an erased nonce) by refraining from sending the corresponding ACK. Consider the following example and assume that the receiver sends back the value of the most-recent nonce instead of a sum: segments 1 and 2 arrive at the receiver, which generally delays its ACKs. CE was set to 1 in segment 1, for example, its nonce of segment 1 was erased. Segment 2 did not experience congestion. Then, a malicious receiver does not even have to go through the trouble of trying to guess what the original nonce value of segment 1 was – all it does is follow its regular procedure of sending an ACK that acknowledges reception of both segment 1 and segment 2. Since an XOR sum reflects the combined value of the two nonce bits, a receiver cannot simply ignore such intermediate congestion events.

A problem with the sum is that a congestion event (which clears the nonce) introduces a permanent error – that is, since all subsequent nonce sums depend on the current value of the sum, a nonce failure (which does not indicate a malicious receiver but only reflects that congestion has occurred) will not vanish. As long as no additional nonces are lost, the difference between the nonce sum expected by the sender and the sum that the receiver calculates is constant; this means that it can be resynchronized by having the sender set its sum to that of the receiver. RFC 3540 achieves this by specifying that the sender suspends checking the nonce as long as it sets CWR = 1 and resets its nonce sum to the sum of the receiver when the next ACK for new data arrives. This requires no additional signalling or other explicit involvement of the receiver – the sender simply takes care of synchronization while the receiver keeps following its standard rules for calculating and reflecting the nonce sum.

Notably, the ECN nonce does not only disable the receiver from hiding a CE mark, but it also has the nice side effect of preventing the 'optimistic ACKing' attack that was described in Section 3.5, as the sender generally does not accept any ACKs that do not contain a proper nonce. Moreover, the receiver is not the only device that would have an incentive to remove ECN marks – the nonce also offers protection from middleboxes such as NAT boxes, firewalls or QoS bandwidth shapers (see Section 5.3.1) that might want to do the same (or do so because they are buggy). The ECN nonce provides quite good protection: while there is always a fifty-fifty chance of guessing the correct nonce in a single packet, it becomes quite unlikely for a malicious user to always guess it in a long series of packets. What exactly a sender should do when it detects malicious behaviour is an open question. This is a matter of policy, and RFC 3540 only suggests a couple of things that a sender could do under such circumstances: it could rate limit the connection, or simply set both ECT and CE to 0 in all subsequent packets and thereby disable ECN, which means that even ECN-capable routers will drop packets in the presence of congestion.

4.1.5 Spurious timeouts

Sometimes, network effects such as 'route flapping' (quickly changing network paths), connection handover in mobile networks or link layer error recovery in wireless networks can cause a sudden delay spike. With the rather aggressive fine-grain timers recommended in RFC 2988, this can lead to expiry of the RTO timer, which means that the sender enters slow-start mode and begins to retransmit a series of segments. Here, the underlying assumption is that the 'pipe' has emptied, for example, there are no more segments in flight. If this is not the case, the timeout is spurious, and entering a slow-start phase that exponentially increases *cwnd* violates the 'conservation of packets' principle. Since timeouts are regarded as a 'last resort' for severe cases as they are generally known to lead to inefficient behaviour, avoiding or at least recovering from spurious ones is an important goal. But first, a spurious timeout must be *detected*.

The Eifel detection algorithm

This is what the *Eifel detection algorithm* does. This simple yet highly advantageous idea, which was originally described in (Ludwig and Katz 2000) and specified in RFC 3522 (Ludwig and Meyer 2003), lets a TCP sender detect whether a timeout was unnecessary by eliminating the retransmission ambiguity problem, which we already encountered in the context of RTO calculation in Section 3.3.1. Figure 4.1 shows how it comes about: assume that a TCP sender transmits segments 1 to 5, and because of a sudden delay spike, a timeout occurs for segment 1 even though all the segments actually reach the receiver. Then, the TCP sender retransmits the segment and, after a while, an ACK that acknowledges reception of segment 1 (shown as 'ACK 2' in the figure) arrives. At this point, there are two possibilities:

1. The ACK acknowledges reception of the retransmitted segment number 1. Everything is all right.

2. The ACK acknowledges reception of the original segment 1 and the timeout was spurious.

By default, a TCP sender cannot distinguish between these two cases. Note that while the timeout event is the more-important one (because the reaction is more severe) the same problem can occur with fast retransmit/fast recovery – that is, the first full ACK that brings the sender back into congestion-avoidance mode is assumed to stem from a retransmitted segment, but this does not necessarily have to be correct; the three consecutive DupACKs that are necessary for the sender to enter loss recovery could also be caused by severe reordering in the network.

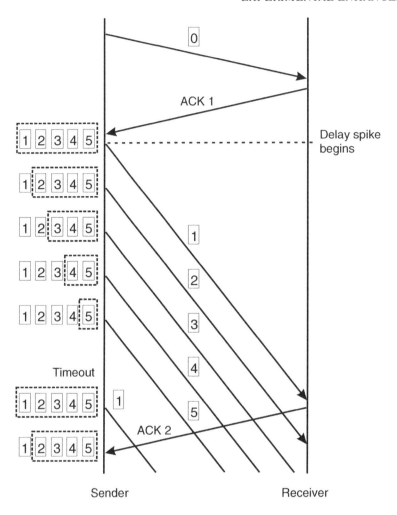

Figure 4.1   A spurious timeout

Eifel solves this problem by mandating usage of the Timestamps option and using the additional information it contains: upon entering loss recovery, the timestamp of the first retransmitted segment (the only one, in case of a fast retransmit without SACK) is stored. This value is compared with the timestamp of the first ACK for new data. If this timestamp is smaller than the stored one, it is clear that this ACK must have been caused by the original segment and not its retransmitted copy, and the loss-recovery phase is detected to be unnecessary. As already discussed in Section 3.3.2, using the Timestamps option is not entirely without cost, as it causes a 12-byte overhead in every payload packet and is not supported by the header compression scheme specified in RFC 1144 (Jacobson 1990). As an alternative, one could use a single bit in the TCP header to distinguish an original segment from its retransmission (Ludwig and Katz 2000) – but there are only few bits left in the TCP header, and this variant was therefore abandoned in the IETF and not included in RFC 3522.

Detection with D-SACK

Theoretically, the retransmission ambiguity problem is also solved by the D-SACK option that was specified in RFC 2883 (Floyd et al. 2000b) because this option explicitly informs the sender that duplicate segments were received. However, it is quite late – D-SACKs are sent when the duplicate segment arrives, and when a D-SACK arrives back at the sender, Eifel would have already detected that the loss-recovery phase is spurious. This being said, using D-SACK for this detection is possible, and (Blanton and Allman 2002) describes how the 'scoreboard' mechanism from RFC 3517 (see Section 3.4.8) can be appropriately extended. This method, which was later specified in RFC 3708 (Blanton and Allman 2004), is slower than Eifel, but it has the advantage of not requiring usage of the Timestamps option in each and every segment.

D-SACK is additionally used by Eifel to cope with one particular corner case: say, a window of 10 segments is sent, all of them reach the receiver and all the ACKs are lost. Then, the first retransmitted segment will cause an ACK that carries the timestamp of the last segment that arrived in sequence (segment number 10) – recall that this is mandated by RFC 1323 in order to make the RTO calculation reflect the fact that the receiver may delay its ACKs. Thus, Eifel will detect the timeout as being spurious, which may not be ideal because losing all ACKs *is* a severe congestion event, albeit on the backward path. With D-SACK, however, the sender can detect that this ACK was caused by a duplicate and decide not to declare the timeout spurious.

F-RTO detection

F-RTO (Sarolahti and Kojo 2004), first described in (Sarolahti et al. 2003), is a sender-side mechanism that manages to detect a spurious timeout (but *only* timeouts, not spurious fast retransmit/fast recovery phases, which can be caused by massive reordering inside the network) without requiring any TCP options. Roughly, it works as follows: upon expiry of the RTO timer, slow start is entered and a segment is transmitted as usual. In addition to standard TCP behaviour, however, F-RTO more carefully monitors incoming ACKs. If the first ACK after the first RTO retransmission advances the window, the RTO might be spurious – but it was decided that this information does not provide enough confidence. So, as an additional test, instead of retransmitting the next two segments, two new ones are sent in the next step. If the RTO is not a spurious one, these two segments cause DupACKs; if, however, the next ACK also advances the window, the RTO is declared spurious. This idea of sending new data in support of recovery resembles limited transmit in RFC 3042 (see Section 3.4.6), but its goal is to detect a spurious timeout rather than to increase the chances for successfully entering fast retransmit/fast recovery.

The F-RTO algorithm is, in fact, a little more sophisticated: it applies some changes to the 'recover' variable that is used by NewReno (see Section 3.4.7 in the previous chapter), and a refinement of the algorithm that uses SACK is specified in (Sarolahti and Kojo 2004). SCTP considerations are also discussed in this document – it turns out that applying the proposed solutions for detecting spurious timeouts to SCTP is not as easy as it may seem (Ladha et al. 2004). At the time of writing, (Sarolahti and Kojo 2004) is in the RFC editor queue, which means that it is quite close to becoming an RFC.

Recovering from spurious timeouts

So far, we were only concerned with *detection* of a spurious loss-recovery phase; obviously, when such an event is detected, something should be done about it. The IETF documents RFC 3522, RFC 3708 and (Sarolahti and Kojo 2004) contain discussions about the reaction to spurious loss recovery, but they do not specify what exactly should be done. Clearly, standard TCP behaviour is not appropriate, as it is based on false conclusions. In particular, this concerns retransmissions of segments – if a sender can be sure that these segments already made it to the other end, it is much more reasonable to transmit new data. Also, as already mentioned, executing slow start is inappropriate in the case of a spurious timeout because it is based on the false assumption that the 'pipe' has emptied. While there are some proposals for recovery from such an event in the literature (e.g. (Blanton and Allman 2002), (Sarolahti et al. 2003) and (Gurtov and Ludwig 2003)), the IETF only recently published RFC 4015 (Ludwig and Gurtov 2005), which specifies such behaviour.

RFC 3522 and (Sarolahti and Kojo 2004) have a common 'plug' that can be connected to a response algorithm: the *SpuriousRecovery* variable. This variable should be set according to the type of event that was detected. '0' (representing *FALSE*) means that no spurious loss event was detected; '1' means that a spurious *timeout* was detected, and if a *false fast retransmit* was detected, the variable must be set to the number of DupACKs that have already been received by a TCP sender before the fast retransmit is sent. In RFC 4015, another value is defined: '−1', which means that a spurious retransmit was detected late (i.e. based upon the ACK for the retransmit, as with D-SACK). Using this variable, the document then specifies the *Eifel response algorithm*, which does quite a number of things:

- Before *cwnd* and *ssthresh* are reduced in response to loss, the number of packets in flight (or *ssthresh* if the sender is in slow start) and the estimated RTT are stored.

- If the spurious loss event is a timeout and it is based on an ACK from the original transmit (not the retransmit, as with D-SACK based spurious loss event detection), the transmission continues with previously unsent data (i.e. the Go-back-N process of the timeout is interrupted).

- Unless the last ACK that came in had ECE = 1, the congestion reaction that was carried out during a spurious loss event was unnecessary, and the Eifel response algorithm reverts to the previously stored state much like the congestion window validation algorithm in RFC 2861 (Section 4.1.3). Notably, this includes restoring *cwnd*, which was reduced to one MSS on the timeout, to approximately its original value. It also fixes the problem that implementations of congestion window validation that strictly follow the pseudo-code in RFC 2861 could misinterpret a spurious timeout as a phase where the sender was idle.

- Knowledge of spurious loss events eliminates the retransmission ambiguity problem and therefore renders Karn's algorithm unnecessary. This means that additional information can be used for RTO calculation; moreover, spurious timeouts can result from a delay spike, which is a special situation. On the basis of these facts, an RTO update procedure is described in RFC 4015.

RFC 4015 is a standards track document. While this indicates that spurious loss event detection and recovery may become recommended TCP behaviour in the future, it is probably too early to regard these mechanisms as anything other than experimental. After all, RFC 4015 only specifies how to *react* to spurious loss events – the documents that specify their detection are still experimental, and it is unclear, at the time of writing, which ones will eventually turn into standards track RFCs.

Avoiding spurious timeouts

It is of course better to *avoid* spurious timeouts in the first place than to recover from them; this can only be done by updating RTO calculation. One such proposal, (Ekström and Ludwig 2004), changes the timer that was specified in RFC 2988 by always calculating two RTO values – a 'Short-term History RTO' and a 'Long-term History RTO' – and choosing the larger value. It follows the original suggestion in RFC 889 (Mills 1983) to refrain from using a fixed value for α; the authors point out that the fixed parameter values in (Jacobson 1988) were based on the assumption that a sender only takes one RTO calculation sample per RTT whereas RFC 1323 suggests not to do so. The newly proposed mechanism is called *Peak Hopper (PH-RTO)*, and in (Ekström and Ludwig 2004) it was shown to do a better job at tracking the RTT than the standard RFC 2988 RTO algorithm.

Another proposal describes how to make use of ACKs for segments that were retransmitted even in the absence of the Timestamps option. While Karn's algorithm (see Section 3.3.1) says that such ACKs should not be used for RTO calculation because of the retransmission ambiguity problem, D-SACK changes this situation somewhat, as the information conveyed by D-SACK clearly indicates that a segment was (unnecessarily) retransmitted. In (Zhang et al. 2002), an idea that makes use of both these ACKs is described: when two ACKs arrive, the second of which indicates an unnecessary retransmission via D-SACK, the sender computes the time between sending the original segment and its ACK and the time between retransmitting the segment and the ACK that contains the D-SACK information. Then, the average of these two values is taken; if one of the segments was delayed, this will have an appropriate influence on the calculated RTO.

4.1.6 Reordering

The main theme of (Zhang et al. 2002) is not its RTO calculation enhancement but the general goal of making TCP more robust against reordering. Note that these things are closely related – for instance, both a delay spike and a reordering event can cause a spurious timeout. Nowadays, massive reordering is not a very common scenario, and when it happens, it is often regarded as a defect. For instance, link layer ARQ which compensates for loss typically does not reorder but only delays packets (Sarolahti et al. 2003); certain mechanisms that could greatly enhance performance are not being developed because they have the potential to reorder packets, and this is known to be a major problem for TCP. One example is multi-path routing, where packets are evenly distributed across a set of paths in a stateless manner (i.e. without distinguishing between flows); note that there is some relationship between this technology and our discussion of congestion control and layers in Section 2.14.1. Another example is parallel forwarding and/or switch hardware, which cannot be fully exploited these days because this would lead to reordering. Thus, it seems reasonable to make TCP a bit more robust against such effects.

The main element of TCP that controls its behaviour in the presence of reordering is the parameter that was called *consecutive duplicates* threshold by Van Jacobson in his email that introduced fast retransmit/fast recovery (see Section 3.4.5). This threshold was specified to be set to 3 in RFC 2581, and this is the value that was used ever since; notably, RFC 3517 treats it as a variable (called *DupThresh*), again in anticipation of future work even though it is still specified to have the same value. *DupThresh* represents a trade-off between robustness against reordering on the one hand and rapidness of response to congestion on the other: for example, setting it to 10 would mean that TCP would wait for 10 DupACKs to arrive before entering fast retransmit/fast recovery. Not only does this delay the congestion response, but it also increases the chance of having the RTO timer expire in the meantime or not even obtaining enough DupACKs to enter fast retransmit/fast recovery at all. On the other hand, setting it to one would mean that even a single packet that overtakes another already causes a congestion response. Thus, while it seems inevitable to tune this parameter in order to make TCP more robust against reordering, finding the right value is not easy.

Changing *DupThresh* also has interactions with limited transmit, which makes the sender transmit new data segments in response to the first two incoming DupACKs in order to keep the ACK clock in motion and increase the chance of obtaining enough DupACKs for triggering fast retransmit/fast recovery (see Section 3.4.6 or RFC 3042 (Allman et al. 2001)). Despite all its advantages, limited transmit is somewhat critical as it sends segments beyond the current congestion window. It would make sense to allow for transmission of (*DupThresh* − 1) instead of exactly two segments with limited transmit when *DupThresh* is changed – according to (Deepak et al. 2001), this should not seriously endanger the stability of the network because the ACK clock is still maintained. This is clearly a change that requires some caution. As such, it should impose an upper limit on the value of *DupThresh*.

The mechanism described in (Zhang et al. 2002) utilizes the scoreboard data structure that was specified in RFC 3517 (Section 3.4.8) to determine the 'reordering length' of packets – this is the number of out-of-order packets that arrive at the receiver before a packet does. This information is stored in a 'reordering histogram' – a data structure that holds the reordering history for a configurable time period. Packet reordering is detected by monitoring incoming ACKs; if a packet is retransmitted, only a D-SACK arriving in response to its retransmission is taken to reliably indicate reordering. If this retransmission was caused by *DupThresh* DupACKs, it is a false fast retransmit. False fast retransmit can be prevented by setting *DupThresh* such that it equals the desired percentile value in the cumulative distribution of the reordering lengths in the histogram. As an example, if $X\%$ of the reordering lengths consisted of five packets or less, setting *DupThresh* to six will avoid $X\%$ of all reordering events (provided that path reordering properties do not change).

A large *DupThresh* value is less harmful for paths where there is hardly any loss. On lossy paths, however, it should be small in order to avoid the adverse effects described above. For events such as timeouts and false fast retransmits, cost functions are derived in (Zhang et al. 2002) – on the basis of these calculated costs, the value of the aforementioned X (called the *False Fast Retransmit Avoidance (FA)* ratio) is decreased. This is perhaps the most-important difference between this work and (Blanton and Allman 2002), where it was already suggested that *DupThresh* be increased in response to false fast retransmits. The new scheme is called *Reordering-Robust TCP (RR-TCP)*, and it balances *DupThresh* by increasing it when long reordering events occur and decreasing it when it turns out that the value was too large. RR-TCP includes the complementary updated RTO calculation method that was described in the previous section. Simulations in (Zhang et al. 2002) show promising results, including the ability of RR-TCP to effectively avoid timeouts.

4.1.7 Corruption

The range of proposals to enhance the performance of TCP in the presence of corruption is sheerly endless. The underlying problem that all these proposals are trying to solve in one way or another is a misunderstanding: when a packet is lost, TCP generally assumes that this is a sign of congestion, whereas in reality, it sometimes is not. When a mobile user passes by a wall, link noise can occur, which leads to bit errors and causes a checksum to fail. Eventually, the packet cannot be delivered, and TCP assumes that it was dropped as a result of congestion in the network (i.e. because a queue overflowed or an AQM mechanism decided to drop the packet). Thus, the goal is to detect corruption and inform the TCP sender about it so that it can take corrective action instead of decreasing *cwnd* as if congestion had occurred. As with congestion, the type of feedback that can be used to deal with corruption can have two forms:

Implicit: Since their feedback loop is typically shorter than the end-to-end feedback loop, it is quite common for link layers to retransmit frames in the presence of corruption. This method, which is called *link layer ARQ*, is typically carried out with a certain level of persistence (which might even be configurable), which means that the system gives up and decides to drop the whole frame after a timeout or after a certain number of tries (more details can be found in RFC 3366 (Fairhurst and Wood 2002)). Thus, from the perspective of TCP, the segment is either dropped or *delayed* – technologies that detect spurious timeouts (see Section 4.1.5), which generally make TCP more robust in the presence of sudden delay spikes, will therefore lead to better behaviour across such links. Also, while one can never rely on ECN as the only congestion signal because it may be inevitable for a router to drop some packets in the presence of a large amount of traffic, this signal is an explicit *congestion* notification, which means that it is not prone to the same misinterpretation as dropped packets. If one could assume that all routers support ECN, total absence of such explicit congestion signals could implicitly indicate that loss might stem from corruption instead; while it is questionable whether it can be exploited in such a manner in practice (after all, loss can *always* indicate congestion), this is still an interesting facet of ECN. Finally, as could be expected, using SACK can make things better because of its ability to cope with multiple packet drops from a single window (which may not be a rare event if a link is noisy).

The end-to-end capacity of a path is implicitly available information that may not be adversely affected by corruption. That is, if a congestion control scheme estimates this factor, it is less prone to the common misinterpretation of packet loss as a sign of congestion (Krishnan et al. 2004). One example of such a mechanism is described in Section 4.6. In general, the greater the reliance on packet loss as the sole means to convey a congestion notification, the greater the chance of misinterpretations.

Explicit: Because the link layer devices adjacent to a link where corruption occurs usually know about the problem (because this is where a checksum fails), one can envision schemes that make use of this explicitly available knowledge, for example, by querying them in a way that resembles our description of explicit rate feedback in Section 2.12.2 (this was called *Explicit Transport Error Notification (ETEN)* and proposed in (Krishnan et al. 2004), and it could be done with a signalling protocol like PTP (see Section 4.6.4)) or having them generate feedback in a way that resembles our description of choke packets in the same section (this was proposed in the context of an IETF effort called *Triggers for Transport* (TrigTran)[2] by the name *Corruption Experienced*, but TrigTran never gained ground). Finally, corruption itself can serve as an explicit signal – that is, if corrupt data are *not* dropped or retransmitted by the link layer devices and such data actually make it to the transport layer instance of the receiver, this is where a checksum will fail and the problem can be noticed.

(Krishnan et al. 2004) contains what may be the clearest and most-thorough overview of mechanisms that strive to enhance IP-based communication across noisy links that was written to date. In this paper, there is a specific focus on mechanisms that utilize explicit feedback, and it contains a classification based on the type of feedback used, granularity (per-path, per-node, per-link, per-flow in case of cumulative feedback, or per-packet feedback), direction, locus and several more properties. Moreover, it contains a description of simulations where TCP SACK was enhanced with an 'Oracle' – an ideal feedback mechanism that would always notify the TCP sender with unrealistically precise information; the results show exactly how big the potential performance improvement from such mechanisms really is. The fact that the inevitable timeout that occurs when a large number of packets is lost imposes a natural upper limit on the benefits is one of the findings.

Many of the mechanisms described in this chapter are useful for noisy links in one way or another. There are also transparent solutions that can be installed *inside* the network without requiring any changes to the TCP specification – but once again, these are technologies that may work particularly well in the presence of noisy links, but they may also be useful in other environments. We will look at them in detail in Section 4.3.1; see Table 4.3 for an overview of the other related schemes dealt with in this chapter. Before we head on to the next topic, let us now take a closer look at one particular suggestion that uses explicit feedback in order to improve the performance of TCP.

Separate checksums

When a bit error occurs, neither UDP nor TCP can recover any reasonable information because the checksums of both these protocols cover the whole transport layer packet (and some IP fields too, turning the header into a 'pseudo header', but we can neglect this detail here). This means that even the port numbers may be erroneous, rendering the whole packet useless; it is neither possible to reuse the corrupt data somehow when a UDP implementation does not even know which application to address, nor can a TCP sender be notified of a corruption event. For this reason, it makes no sense for a link layer device to forward such data – since the link layer control loop is typically shorter than the end-to-end

---

[2]http://www.ietf.org/mailman/listinfo/trigtran

control loop, it is more efficient to either retransmit locally or drop a corrupted frame at this point.

The *Lightweight User Datagram Protocol (UDP-Lite)* changed things somewhat. Its specification, RFC 3828 (Larzon et al. 2004), altered the meaning of the redundant UDP 'length' field to 'checksum coverage' and mandated that the standard checksum of this new protocol only cover as much as specified in this field. The reason for doing so is that there are video and audio codecs that are designed to cope with bit errors, but if erroneous data arrive at the transport layer, UDP always drops the whole packet even if only a single bit is wrong. With normal UDP, it is therefore hardly possible to exploit these codec features (unless very small packets are used, which has the disadvantage of increasing the relative amount of bandwidth wasted for headers).

UDP-Lite caused quite a long series of discussions in the IETF. Simple as it is, it took more than two years from its first proposed form in an Internet-draft until it was actually published as a standards track RFC. Its criticism included the fact that it would now be possible to write a TCP/IP based application that does not work in one environment (where there are lots of bit errors) and seamlessly works in another. This dependence on lower layer behaviour can be interpreted as going against the fundamental notion of 'IP over everything' and thereby breaking the basic Internet model. Finally, a decision was made to specify that link layers should always check whether a packet is corrupt and only forward known-corrupt packets if they are known to be UDP-Lite packets. Requiring link layer devices to notice transport headers is a pretty ugly layer violation, but it may be the most prudent choice given that there is no better means[3] available for inter-layer communication. In any case, it might now be regarded as acceptable to assume that some future link layers could detect a transport header and use it to decide whether forwarding known-corrupt data is appropriate.

Under these circumstances, it seems that the same could be done with TCP. Here, applications are generally not interested in corrupt data (the combination of 'reliable delivery' and 'corrupt' is perhaps not a happy one), but detecting that corruption occurred and informing the sender of this event may be quite beneficial. This was specified in (Welzl 2004), which is based upon *TCP HACK* (Balan et al. 2001). Here, the idea is to have packets carry a 'Corruption Detection' option that contains an additional checksum; this second checksum only covers the relevant parts of the header,[4] and if the regular checksum fails but the checksum in the option does not, it can unambiguously be deduced that the packet has arrived but was corrupted. Specifically, the 'pseudo header' used by this checksum preserves the ECN field, source and destination address of the IP header, source and destination port, the sequence and ACK numbers and the CWR, ECE, ACK, RST, SYN and FIN flags as well as all options. Some parts of the header are irrelevant; for instance, since the data will not be delivered if the checksum fails, the urgent pointer, URG and PSH flags need not be covered by this checksum, and making the checked part as small as possible reduces the risk of corruption occurring therein.

A second option called *Corruption Notification* is used to convey the information 'segment number X was received, but its contents were erroneous' to the sender. The

---

[3]One could imagine a 'Corruption Acceptable' bit in the IP header, but the IP (and especially the IPv6) header hardly has any bits to spare.

[4]This was suggested by Craig Partridge as an alternative to covering the complete header.

specification does not prescribe what a sender should do with this information; it only provides a means to detect corruption and carry out the necessary signalling. There are several possibilities:

- It would seem logical not to reduce *cwnd* as if congestion had occurred. However, simply maintaining *cwnd* as if there was no congestion whatsoever may be too imprudent – it has been said that some congestion may manifest as corruption, for example, on shared wireless links.

- If the information from the receiver is 'corruption has occurred, and CE was set to one', then it would make sense for the sender to reduce *cwnd* right away instead of waiting for the next ACK or a timeout. This way, congestion can be detected earlier, thereby decreasing the chance of sending too much and unnecessarily having packets dropped.

- The sender could retransmit the corrupt segment when the next transmission is scheduled instead of waiting for a loss notification in the form of DupACKs or a timeout.

- If the regular checksum fails but the checksum in the 'Corruption Detection' option does not, and the packet is an ACK that does not contain any data (e.g. when TCP is used in a unidirectional manner), we have quite an advantageous scenario. This ACK successfully conveys the information 'segment number X was received', and the fact that some header fields were erroneous does not necessarily indicate any problem along the forward path. This advantage is gained without the necessity for signalling any related information to the other end of the connection. However, since ACKs are typically small, the chances for corruption to occur in the sensitive part of the header (the pseudo header used for the checksum in the Corruption Detection option) are high, and so is the relative header overhead of including the option.

- In any case, it would seem reasonable to utilize ACKs for segments that are known to be corrupt or even ACKs that are known to be corrupt for RTO calculation. If a large series of packets experience corruption, the fact that this can significantly impair RTO calculation has quite severe consequences (Krishnan et al. 2004), and therefore, this feature has the potential to be quite advantageous in such a scenario.

More research is required to judge whether the potential benefits from using the 'Corruption Detection' and 'Corruption Notification' options outweigh the cost. Evaluating such a scheme is not easy, as link models in simulations may not always do so well in capturing the behaviour of real links; there is some recent work that may alleviate this problem (Gurtov and Floyd 2004). An overview of the general issues related to handing over known-corrupt data can be found in (Welzl 2005b). Carrying out real-life tests is also not easy because, as already mentioned, most link layers do not readily hand over known-corrupt data nowadays; this work is for future link layer technologies that may be designed to detect transport layer schemes that want to receive data even if bit errors occur.

While it is very questionable whether the IETF will ever accept the integration of separate checksums into TCP, you did not read this section in vain: as we will see in Section 4.5.2, the very same feature was integrated in another protocol.

4.2 Maintaining congestion state

Congestion occurs along a network path – between two IP addresses. Since TCP is a transport layer protocol, several instances of the protocol with different port numbers can communicate across the same channel. In other words, computer A may run an FTP server, and computer B may run several FTP clients that download files from computer A at the same time. The parallel FTP (and hence, TCP) connections happily live side by side and do not notice or care about each other – yet, they share the same network path with similar congestion properties. The following proposals are concerned with managing such a shared congestion state.

4.2.1 TCP Control Block Interdependence

TCP states such as the current RTT estimate, the scoreboard from RFC 3517 (Blanton et al. 2003) and all other related things are usually stored in a data structure that is called the *TCP Control Block* (TCB). Normally, this block is implemented once for each connection – every TCP instance has its own independent TCB. RFC 2140 (Touch 1997) calls for a more-open approach when initializing the TCB. Note that the suggestions in this document are only related to the initialization phase; states that are updated when a connection is already active remain unchanged by them, and therefore RFC 2140 has a minor impact on the long-term behaviour of a connection.

Still, the initial case is an especially interesting one, as this is a situation where a TCP sender is normally assumed to have no idea at all about the network state. This concerns, for instance, the MSS[5] and the most-recent RTT estimate. RFC 2140 distinguishes between two types of sharing: *temporal* sharing, where a TCB of a TCP connection that is already terminated is used, and *ensemble* sharing, where a number of TCP flows are still active while their state is used to initialize the TCB of a newly started one. In such a scenario, one might want to benefit from the readily available *cwnd* of other flows, but the question of how to ideally share such information is a tricky one – should the congestion window of a new flow be initialized with $cwnd/n$, where n is the total number of flows that share a path? Such questions are tackled in (Eggert et al. 2000), where it is shown that TCB sharing is beneficial for access to web sites with HTTP 1.0 (which opens a new TCP connection to download each and every object contained in a web page[6]). Some more performance results of TCB sharing can be found in (Savorić et al. 2003).

4.2.2 The Congestion Manager

This concept was taken to the next level in (Balakrishnan et al. 1999), which describes a *Congestion Manager (CM)* – a single per-host entity that maintains all the states required to carry out congestion control and provides any flows with fully dynamic state sharing capabilities. Figure 4.2 shows how it works: instead of solely maintaining variables of their

---

[5]Actually, the MSS is a special case because it should depend on the MTU of a path, and RFC 1191 (Mogul and Deering 1990) already prescribes such shared usage of the information collected with the 'Path MTU Discovery' mechanisms.

[6]This problem was solved by introducing *persistent connections* in HTTP 1.1 as specified in RFC 2616 (Fielding et al. 1999). Note that this is more reasonable from the network stability point of view, but HTTP 1.0 may be more efficient for the user if several TCP connections are used in parallel.

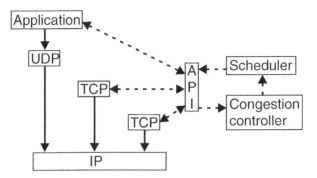

Figure 4.2 The congestion manager

own, TCP instances query a common CM API for the current value of *cwnd* (the broken lines indicate the control flow) and inform the CM of updates via an 'update' function in the same API. This effectively turns *n* simultaneously operating congestion control instances per path (such as the multiple FTP connections that were mentioned in the beginning of this section) into a single one – the scenario of one instance per path is what the congestion control algorithms in TCP were designed for.

Figure 4.2 shows another important aspect of the congestion manager: besides the TCP instances, an application that utilizes UDP to transfer its data also queries the API. Such applications should take care of congestion control themselves, or they endanger the stability of the Internet (Floyd and Fall 1999). This is easier said than done: not only is the application level the wrong place for congestion control (precise timers may be required), implementing it is also an exceedingly difficult task. Here, the CM can help, as it offloads all this functionality into the operating system (where it belongs), enabling an application to simply query for the calculated information in order to better adapt its rate.

The CM as described in (Balakrishnan et al. 1999) consists of changes both on the sender and the receiver side, and uses its own protocol header for signalling between the two. The receiver is necessary for providing feedback if the application cannot do so; while the CM is shown to be beneficial even if only the sender-side changes are applied, this requires UDP-based applications to take care of feedback signalling themselves and refresh the state of the CM by calling the 'update' function in the API whenever feedback arrives. The sender realizes a window-based AIMD scheme with additional rate shaping (making it a hybrid window-based/rate-based congestion control mechanism).

RFC 3124 (Balakrishnan and Seshan 2001) is the IETF specification of the CM. It consists of only the sender part – the receiver functionality as well as the header described in (Balakrishnan et al. 1999) are not included in this document. The API provides for two types of transmission modes: *callback-based*, where the application first calls a 'request' procedure that informs the CM that it desires to transmit data, and then a callback procedure is invoked in order to send segments exactly when the CM decides to do so, and *synchronous-style*, where the application simply calls a function that returns the rate that it should use at this time. The latter mode was included for ease of implementation. Since callbacks provide for the same functionality, procedure calls for buffered congestion-controlled transmissions were intentionally omitted from the API.

Additional API functions include the aforementioned 'update' procedure, which lets applications inform the CM about events such as congestion losses, successful receptions, RTT samples, ECN marks and so on. All these things can be deduced from feedback messages such as TCP ACKs. Whenever an application actually transmits data, it must call a 'notify' procedure, which enables the CM to keep track of when packets are sent. Finally, when the callback-based transmission mode is used, the callback function receives an expiration time, which must not be exceeded before transmitting a packet – otherwise, the congestion control behaviour dictated by the CM would be distorted by the application. Internally, the main congestion controller communicates with a *scheduler* (shown in Figure 4.2); this is where flow multiplexing is taken care of, and decisions are made regarding which flow is allowed to send when.

Deployment of the CM is a critical issue. RFC 3124 should not even be in this chapter, as it is a standards track protocol; yet, (Duke et al. 2004) states:

> Although a Proposed Standard, some pieces of the Congestion Manager support architecture have not been specified yet, and it has not achieved use or implementation beyond experimental stacks.

and a recent query about CM deployment in the IRTF end-to-end interest mailing list yielded no response. This does not mean that such mechanisms have no chance for deployment whatsoever – (Duke et al. 2004) also mentions that TCB sharing as described in RFC 2140 is at least partially done in practice by a few operating systems (e.g. Linux has a destination cache). Other than simpler TCB variants that do not dynamically share congestion state, the CM may also suffer from an incentive problem: since several flows that share a single path are more aggressive without the CM than they would be with it, it is difficult to see what should convince an operating system designer to implement this rather complex piece of software. These problems resemble potential future deployment issues of the 'Datagram Congestion Control Protocol'; we will discuss further details in Section 6.2.3.

4.2.3 MulTCP

In 1998, when the idea of providing differently priced service levels (QoS) to Internet users was still quite a hype, (Crowcroft and Oechslin 1998) was published. In this paper, the authors describe a protocol called *MulTCP*. Its underlying idea is pretty much the opposite of the congestion manager: instead of trying to group several transport level congestion control instances into together, the fact that several TCP streams are more aggressive than one is exploited. For instance, a flow that stems from a high-class user who pays more than most others could be allowed to constantly act like two TCP senders instead of one. Basically, MulTCP provides this functionality within a single protocol. The underlying idea is that tuning TCP in such a manner can lead to a weighted proportionally fair network utilization, where the utility functions of all users are maximized. This is driven by the finding that AIMD can realize proportional fairness, which was documented in (Kelly et al. 1998); note that MulTCP was developed two years before (Vojnovic et al. 2000) was published – this paper made it clear that this result does not apply to TCP.

Obviously, acting precisely like an arbitrary number of TCP flows at the same time is not reasonable; the control must be constrained somewhat. For instance, if 10 flows would carry out their slow-start phase at exactly the same time, this would mean that the sender would start by transmitting 10 packets (or 20, if the larger initial window specified in RFC 3390 (Allman et al. 2002) is used – in cases where the MSS is small, RFC 3390 even allows an initial window of *four* segments, which would increase the number in our example to 40 packets) into the network before executing any kind of control. In order to prevent such bursts of traffic, MulTCP starts with only one packet and increases its rate by sending three (instead of two) packets for each acknowledgement received in slow-start mode. This is not necessarily slower, but it is smoother than the behaviour of *n* normal TCP senders that send two packets for each acknowledgement received; provided that slow start does not end earlier because *cwnd* exceeds *ssthresh*, MulTCP will have the same window that *n* standard TCP senders would have after a while. Then, it continues its operation just like normal TCP would, by sending two packets for each acknowledgement received. After *k* RTTs, *n* TCP senders have a congestion window of $n * 2^k$ whereas MulTCP has a congestion window of 3^k; these functions yield the same values when *k* equals $\frac{\log n}{\log 3 - \log 2}$. This means that MulTCP simply has to compare its window with three to the power of this fraction whenever it receives an ACK, and switch from its more aggressive increase mode to standard TCP behaviour if this value is exceeded.

In addition to slow start, the behaviour during congestion avoidance and timeout reaction is discussed in (Crowcroft and Oechslin 1998). Emulating the linear increase control of *n* flows is straightforward: the window is increased by $n/cwnd$ instead of $1/cwnd$ for every ACK that is received. The multiplicative decrease step that is carried out upon the reception of three DupACKs is slightly different: if only a single packet is dropped, only one out of *n* TCP flows would react. Since *cwnd* is the sum of the congestion window of *n* virtual senders, only one of these *cwnd* parts would need to be halved – thus, MulTCP sets *cwnd* and *ssthresh* to ($cwnd * \frac{n-0.5}{n}$). When a timeout occurs, *ssthresh* is also set to this value; according to (Crowcroft and Oechslin 1998), there is not much more that can be done to cope with the fact that MulTCP is simply more prone to timeout than *n* individual TCPs are. This is due to the fact that MulTCP realizes a single control loop rather than *n* individual ones, where bursts of packet losses would be distributed across *n* connections. In the latter case, it is unlikely for such a loss event to cause enough packet drops from a single connection to trigger a timeout.

This timeout-related problem limited the efficiency of MulTCP in simulations; if *n*, the number of emulated flows, was greater than 2, the throughput gain did not go above 2.5 (which means that it did not exceed 2.5 times the throughput of a single TCP connection) when MulTCP based on TCP Tahoe, Reno or NewReno was used. Only SACK-based MulTCP, which manages to efficiently recover from multiple dropped packets from a single window and therefore does a better job at avoiding timeouts job than the other TCP flavours do, managed to significantly increase its throughput gain beyond 2.5 as *n* grew. An additional study of MulTCP led to the conclusion that *n* should not be greater than four in order to maintain predictable behaviour (Gevros et al. 1999).

MulTCP is surely an interesting approach at service differentiation; it could be used to realize a QoS scheme of sorts without requiring additional help from routers as with the standard QoS mechanisms described in Section 5.3. This is in line with the pricing-based approaches such as the European M3I project, which we have discussed in Section 2.16.3. While its implementation is rather simple and it can be deployed without changing software in routers, MulTCP was not brought to the IETF and is unlikely to become widely used even in the future; there are now more-modern approaches that exhibit behaviour that is more aggressive than standard TCP. Yet, this protocol played its role in the history of experimental congestion control approaches, and it is an interesting counterpart to the CM.

4.3   Transparent TCP improvements

It is possible to enhance the performance of TCP without changing the standard or even its implementation. This can be carried out, for instance, by placing a device inside the network that does some things to a TCP flow without having any of the end systems (or users) notice its existence. Improvements could also be carried out at the sender by adapting the way TCP is used – in this case, this would not have to be known to the receiver. As we have already seen in Section 3.5, the other direction is also possible: a receiver can increase the throughput that it obtains by 'fooling' a sender. This, however, is generally regarded as an attack because it makes the sender deviate from the desired behaviour and can therefore harm the stability of the network. In this section, we will look at some performance enhancements that are transparent in one way or another without severely violating the general congestion control principles of TCP.

4.3.1   Performance Enhancing Proxies (PEPs)

According to RFC 3135 (Border et al. 2001), the IETF calls any intermediate network device[7] that is used to improve the performance of Internet protocols on network paths where performance suffers because of special link characteristics as a *Performance Enhancing Proxy* (PEP). This can include application layer devices such as web caches. Actually, a PEP can operate at almost any network layer or even distribute its functionality across several layers and protocols at the same time; since PEPs are often transparent to end systems and hiding specifics regarding their operation may yield a business advantage for the company that builds and sells them, PEPs might do things that we can only guess. Luckily, there are some old and well-known tricks that we can look at, and chances are that they are still in use in some of the PEPs out there. In this section, two of these transport level PEP mechanisms for TCP will be discussed on the basis of their descriptions in RFC 3135.

Figure 4.3 shows a scenario where introducing a PEP makes sense: payload packets are sent from left to right, across a wireless link and a satellite. Under normal circumstances, ACKs go back all the way from the receiver to the sender; this is indicated by the backward arrow labelled '1' in the figure. For our observations, it is of minor relevance whether the backward path is the same one as the forward path (i.e. the network is symmetric) or not. In this example, TCP would react rather slowly to changes in network conditions because the RTT is large – in congestion avoidance, the sender increases its rate by at most one segment per RTT. If packet drops from corruption are frequent along the wireless link, chances are that the sender can hardly fully recover before the next drop occurs, and the resulting overall link utilization could be very low.

---

[7]We will regard PEPs as special devices here, but this can of course also be a function that is implemented in a standard router which does many other things at the same time.

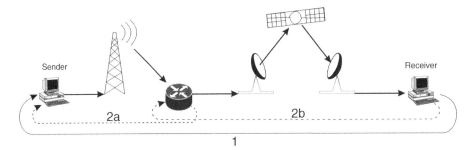

Figure 4.3 Connection splitting

Connection splitting

One method to alleviate this problem is *connection splitting*; a well-known (albeit old) realization of this mechanism is called *Indirect-TCP (I-TCP)*. (Bakre and Badrinath 1995). In our example scenario, connection splitting works as follows: a PEP is installed at the router that lies between the wireless link and the left satellite base station. This device now acts like a regular TCP receiver towards the sender – it acknowledges packets and buffers them. In addition, it acts like a regular TCP sender towards the receiver – it transmits the buffered packets that came from the original sender and waits for acknowledgements. In order to be fully transparent to the end systems, the PEP would have to use the address of the receiver and the sender, respectively; this is known as *TCP spoofing*.

The advantages of such an approach are obvious: since there are now two separate control loops ('2a' and '2b' in the figure), each of which has a shorter RTT, error recovery can work better. If the wireless link is noisy and therefore frequently causes packet drops, the sender will now notice the problem much faster because the PEP right after the receiving end of the wireless link will immediately start sending DupACKs. This could be taken one step further by installing an additional PEP in the left and right satellite base station. There may be several reasons for doing so: first, satellite links are also known to be noisy, which might make shortening the control loop seem to be a good idea (on the other hand, this is often compensated for with FEC). Second, TCP is known to be a poor match for long fat pipes, and satellite links happen to have this property. This is pronounced when the TCP implementation is outdated (e.g. it does not support the 'Window Scale' option described in Section 3.4.8) – using a more-suitable TCP variant across the satellite can significantly enhance the performance in such cases. Later in this chapter, we will discuss a number of protocols that were designed exactly for such environments, and PEPs in the base stations could be used to transparently switch from TCP to a more-suitable protocol and back again.

There are also some issues with connection splitting, which is mainly due to the fact that this method is at odds with the end-to-end argument (see Section 2.11.1):

- It violates the end-to-end semantics of TCP. When a TCP sender receives a cumulative ACK, it assumes that all the bytes up to (but not including) the number in the ACK were correctly received by the other end. This is not the case when a PEP makes the sender believe that it is the receiver.

- The end-to-end argument encompasses an important principle known as *fate sharing*, which is related to the connection-specific state that is stored in the network. Ideally, only the communicating endpoints should store the state related to a connection; then, these peers can only disagree on the state if the path in between them is inoperable, in which case the disagreement does not matter. Any additional state stored somewhere in the network should be 'self-healing', that is, robust against failure of any entity involved in the communication. Connection splitting is an example of such additional state that may *not* be self-healing: if the sender wrongly assumes that ACKed packets were correctly received by the other endpoint of the connection, and the intermediate PEP fails for some reason, the packets in its buffer are lost and the connection may end up in an invalid state – the PEP has effectively 'stolen' packets from the connection, but it cannot give them back. Another way of saying this would be: connection splitting makes the end-to-end connection somewhat less reliable.

- Breaking the end-to-end semantics of a connection also means that end-to-end security cannot prevail, and it does not work with IPSec.

- Finally, connection splitting schemes can have significant processing overhead; the efficient maintenance of the intermediate buffer in the face of two asynchronously operating control loops may not be an easy task. For example, loop '2a' in Figure 4.3 could fill the buffer in the PEP much faster than loop '2b' would be able to drain it – then, the sender must be slowed down by some means, for example, by advertising a smaller receiver window. In the meantime, the congestion window of control loop '2b' could have grown, and all of a sudden, the PEP might be required to transfer a large amount of data – it must strike a balance here, and the fact that such devices should typically support a large number of flows at the same time while maintaining high throughput does not make the task easier. Several research endeavours on fine-tuning PEPs have been carried out; one example that tries to preserve ACK clocking across two split connections is (Bosau 2005).

Snoop

Snoop (sometimes called Snoop TCP or Snoop protocol) (Balakrishnan et al. 1995) is quite a different approach; here, the PEP does not split a connection but carries out a more-subtle form of control instead. Most importantly, the end-to-end semantics of TCP are preserved. The Snoop agent monitors headers of packets flowing in both directions and maintains soft intermediate state by storing copies of data packets. Then, if it notices that a packet is lost (because the receiver begins to send DupACKs), it does not wait for the sender to retransmit but does so directly from its buffer. Moreover, the corresponding DupACKs are suppressed; here, it is hoped that the retransmitted packet will cause a cumulative ACK that will reach the 'real' TCP sender before its RTO timer expires, and the loss event is hidden from it. All this merely looks like somewhat weird network behaviour to the TCP endpoints – from their perspective, there is no difference between a PEP that retransmits a dropped packet and a packet that was significantly delayed inside the network, and there is also no difference between a DupACK that was dropped by a PEP and a DupACK that was dropped by a regular router because of congestion. TCP was designed to cope with such situations.

Despite its advantages, Snoop is a poor match for a scenario such as the one depicted in Figure 4.3, where DupACKs travel a long way, and they may signify congestion somewhere along the path; trying to hide such an event from the sender can make the situation worse. Also, the receiver is ideally connected to the wireless link, and the Snoop agent is connected to its other end. RFC 3135 describes variants of Snoop that take into account scenarios where the sender is connected to the wireless link; the Snoop agent can, for instance, send an explicit loss notification of sorts to the sender or use SACK to notify it of a hole in the sequence space. Clearly, the idea of monitoring a TCP connection and intelligently interfering in one way or another allows for a large diversity of things that can be done; as another example, a Snoop agent could refrain from suppressing the first two DupACKs if a sender realizes limited transmit as specified in RFC 3042 (Allman et al. 2001) (see Section 3.4.6 – the idea is to send new data in response to the first two DupACKs) in order to keep the ACK clock in motion.

The two types of PEPs that we have discussed are primarily designed for wireless links, although they may be able to improve the performance in other scenarios too. These are by no means all the things that could be done (e.g. a PEP does not even have to restrict its operation to TCP), and there are many other scenarios where a PEP can be useful. One example is a 'VSAT' network, where a central hub transfers data to end systems across a satellite and data flows back to the hub using some other technology. The topology of such a network is a star; if one endpoint wants to communicate with another, it must first contact the hub, which forwards the data to the receiver via satellite. Such architectures are normally highly asymmetric – the bandwidth from the hub to the receivers is greater than the bandwidth along the backward path, which is normally constrained by the capacity of a terrestrial modem. According to RFC 3135, 'VSAT' PEPs often encompass various functions and typically realize a split connection approach.

In the next section, we will take a look at a function that is also particularly beneficial across satellite links. It can also be realized as a PEP; since PEPs are usually associated with middlebox functions that at least resemble connection splitting or Snoop in one way or another, and this mechanism is entirely different, it deserves a section of its own.

4.3.2 Pacing

As we have already seen in Section 2.7, congestion can cause packets and their corresponding ACKs to be irregularly spaced; this is illustrated in Figure 2.8. In addition, TCP is bursty by nature. Consider, for example, a connection that is in slow-start mode and traverses a long-delay path. The sender will start with one segment, wait for an ACK, send two, wait for the ACKs and so on. For the network, this is a continuously growing burst of packets with one RTT in between. While Van Jacobson considered a state of equilibrium where a somewhat regular stream of packets is sent and the corresponding incoming ACKs continuously clock out new packets in (Jacobson 1988), he also had to design the slow-start mechanism as a means to start the 'ACK clock'. During slow start, ACK clocking does not work in its ideally envisioned manner, and this can lead to such bursts. Irregular spacing is not at odds with the TCP congestion control algorithms, which are primarily concerned with the number of packets per RTT – but this is not the timescale of the network. A TCP sender that transmits all of its packets during the first half of its RTT can cause transient queue growth.

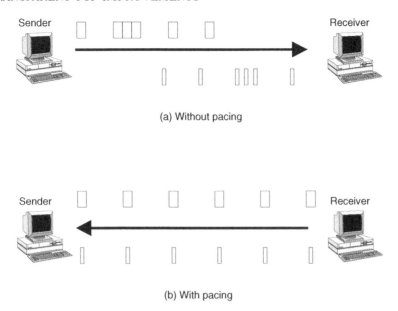

(a) Without pacing

(b) With pacing

Figure 4.4 Pacing

While the inherent burstiness of TCP did not appear to cause significant problems for a long time, the increasing capacity of links used in recent years changed this situation; a desire to restrain the bursts arose. One mechanism that does so is limited slow start (see Section 4.1.2 and RFC 3742 (Floyd 2004)), but there is also a much simpler method: since the related RFCs only provide *upper* limits on the amount of data that a sender can transmit at a given time, it fully conforms with the specification to simply delay packets. This is called *pacing* (or 'rate-based pacing'); the goal is to do this in such a way that the stream changes from something like (a) in Figure 4.4 to something like (b) in the same figure. Packets should be equally distributed – ideally, this is attained by transmitting at an exact rate of $window/RTT$, where $window$ is the current effective sender window.

Pacing can obviously be carried out by the sender, but it can also be carried out by the receiver (which only has a choice to delay packets within a certain range) and within the network, which means that the device or piece of software that realizes this functionality is a PEP of sorts. The latter option is particularly attractive not only because of its transparent nature (it does not do any harm to the TCP connection, yet remains completely unnoticed) but also for another reason: pacing is especially important across high-bandwidth networks, where it may not be possible for the end system to generate packets with the desired spacing because its timers are too coarse. For example, in order to equally distribute packets with a standard size of 1500 bytes across a 1 Gbps link, a packet would have to be generated every 11.5 µs (Wei et al. 2005) – a normal PC may not be able to do this, but dedicated hardware might.

In (Takano et al. 2005), this problem is solved in a simple yet effective manner: the sender never waits, but it transmits dummy packets – so-called gap packets – between the actual data packets. The size of these gap packets controls the delay between actual data packets. Since gap packets should not waste bandwidth, they should be discarded by the first hop after the sender; this can be attained by choosing a suitable packet (actually *frame*) type from the underlying link layer technology. For instance, 802.3x defines a PAUSE frame that can be used as a gap packet when its 'pause time' is set to zero.

4.3.3   Tuning parameters on the fly

Another way to control the behaviour of TCP without changing its implementation is to adaptively tune its parameters. One particularly interesting parameter is the buffer size of the sender, which automatically imposes an upper limit on its window. Tuning the maximum window size itself may seem more attractive because its impact is perhaps more immediate, but this parameter is not always available – remember that the goal is to influence TCP without changing its implementation. Since the maximum sender window should ideally equal the bandwidth × RTT product of the connection, the required buffer size varies as widely as the environments that TCP/IP is used in; an example range given in (Hassan and Jain 2004) and (Feng et al. 2003) is a short modem connection with a capacity of 56 kbps and a delay of 5 ms – corresponding with a window size of 36 bytes – versus a long-distance ATM connection with a capacity of 622 Mbps and a delay of 100 ms, which corresponds with a window size of 7.8 MB. Choosing the latter window size for the former scenario wastes over 99% of its allocated memory whereas choosing the former window size for the latter scenario means that up to 99% of the network capacity is wasted.

One simple solution to the problem is to manually tune the buffer size depending on the environment. The problem with this approach is that both the network capacity that is available to an end system and the RTT fluctuate, which is caused by effects such as routing changes or congestion in the network. It is therefore desirable to automatically adapt the buffer size to the given network conditions. While they are not the only methods available (e.g. the memory management technique in Linux kernel version 2.4 also does this), two well-known approaches can be seen as representatives:

1. *Auto-tuning* utilizes TCP header information and the Timestamps option to estimate the bandwidth × RTT product and adapt the buffer size on the fly; this is a sender-side kernel modification where several concurrent TCP connections can share a single buffer (Semke et al. 1998).

2. *Dynamic Right-Sizing (DRS)* is a receiver-side modification that makes the sender change its maximum window by tuning the advertised window (Feng et al. 2003); thus, the *flow control* functionality of TCP is used by this approach. The receiver estimates the current *cwnd* of the sender by monitoring the throughput; the RTT is estimated by measuring the time between sending an ACK and the reception of data that are at least one window beyond the ACKed data. Since this assumes that the sender will transmit new data right away when it receives an ACK, but this may not always be the case (i.e. the sender may not have new data to send), this RTT is interpreted as an upper bound. If the receiver is itself sending back data (remember, TCP connections are bidirectional), the RTT will automatically be estimated by the TCP implementation.

While these techniques do not require changing the TCP code, both auto-tuning and DRS operate at the kernel level. For instance, the advertised receiver window is usually not a parameter that is freely available for tuning to any TCP-based application (note, however, that any changes only have to be carried out at one end of the connection). Since ease of use is perhaps the main goal of parameter tuning – after all, if changing something in the kernel is required, one could also change TCP itself – a user space method to realize DRS is also presented in (Feng et al. 2003) by the name 'drsFTP'. This FTP implementation tunes the receive buffer size, which must somehow affect the advertised window; since ACK sending times and sequence numbers are unknown to applications, the only way to determine the RTT is to send some additional data. In the case of drsFTP, a small packet is sent on the FTP control channel.

While the results presented in (Feng et al. 2003) indicate a significant performance improvement from using drsFTP, user space parameter tuning clearly suffers from the lack of access to information from the transport layer; after all, what is the benefit of hiding information such as the current RTT estimate from applications? The authors of (Mogul et al. 2004) make a point for visibility of such transport data; an idea for taking this kind of information exchange a step further is outlined in Section 6.3.

4.4 Enhancing active queue management

Citing a fundamental similarity between the 'Flyball Regulator' (a device to control steam engines) and the RED control problem, Van Jacobson stated at a NANOG[8] talk that RED will also work with a different control law (drop function) (Jacobson 1998). Specifically, the slides from his talk contain the following statement:

> RED works even when the control law is bizarre. But it works really well when the control law incorporates the additional leverage caused by TCP's congestion avoidance and timeout algorithms.

Taking this fact into account, it is no surprise that researchers have come up with a plethora of proposals to enhance RED.

RED is known to be sensitive to parameter settings, which should ideally depend on the environment (i.e. the nature of the traffic traversing the link under control). Since the environment itself is prone to changes, this can lead to queue oscillations, which are undesirable. One reason to avoid oscillations is that they make the delay experienced by packets somewhat hard to predict – but this is part of the service that an ISP provides its customers with, and therefore it should not underlie unforeseeable fluctuations. Thus, many proposals for RED enhancements focus on stabilizing the queue length. Another common goal is to protect responsive flows, which typically is the same as enforcing fairness among the flows on a link.

While there is no IETF recommendation for any schemes other than RED, which is explicitly recommended in RFC 2309 (Braden et al. 1998), there is also no fundamental issue that could *prevent* router manufacturers from utilizing any of the 'experimental' mechanisms described in this section. AQM is always a complementary operation that does not harm but support TCP; if mechanism X works better than mechanism Y, there is really no reason not to use it. Since their fundamental goals generally do not differ much, even having a diversity of different schemes handle packets along a single path will probably

---

[8]The North American Network Operators' Group.

not do much harm. This fact may render the mechanisms in this section slightly more important than some other things in this chapter, and it may have caused related research to gain momentum in recent years.

It is obviously impossible to cover all the efforts that were made here, especially because some of them delve deeply into the depths of control theory and general mathematical modelling of congestion control (e.g. some of the work by Steven Low and his group – (Srikant 2004) is a much better source for the necessary background of these things). I picked a couple of schemes that I thought to be representative and apologize to authors whose work is equally important yet was not included here. A quite thorough overview and performance evaluation of some more AQM mechanisms can be found in (Hassan and Jain 2004).

Finally, it should be pointed out that all AQM schemes can of course either drop packets or mark them if ECN is available, and ECN always yields a benefit. For simplification, 'marking' and 'dropping' are assumed to have the same meaning in this section, and the 'drop probability' is the same as the probability of marking a packet if ECN is used.

4.4.1 Adaptive RED

As mentioned in Section 3.7, suitably tuning RED is not an easy task. In fact, its parameters should reflect environment conditions for optimal behaviour – the degree of burstiness that one wants to accommodate, for instance, is a direct function of 'typical' RTTs in the network, but such a 'typical' RTT is somewhat hard to determine manually. Ideally, the setting of max_p should even depend on the number of connections, the total bandwidth, segment sizes and RTTs in the network (Hassan and Jain 2004). It is therefore questionable whether having fixed values for RED parameters is a good idea at all – rather, one could carry out measurements and automatically update these values on the fly.

This is the underlying idea of *Adaptive RED*, which was originally described in (Feng et al. 1999): on the basis of the dynamics of the queue length, the max_p parameter is varied. This makes the delay somewhat more predictable because the average queue length is under the control of this parameter. When the network is generally lightly loaded and/or max_p is high, the average queue length is close to min_{th}, and when the network is heavily congested and/or max_p is low, the average queue length is close to max_{th}. Adaptive RED was refined in (Floyd et al. 2001) – this updated version of the algorithm automatically sets other RED parameters, thereby taking additional burden from network administrators. All that needs to be configured is the desired average queue length, which represents a trade-off between utilization and delay.

The only parameter that is altered on the fly is max_p; from (Floyd et al. 2001), the changes to the way that max_p is adapted are as follows:

- The target value is not just between min_{th} and max_{th} but within a range half way between these two parameters.

- max_p is adapted in small steps and slowly (over timescales greater than a typical RTT). This is an important change because it maintains the robustness of the algorithm by allowing the original RED mechanism to dominate the dynamics on smaller timescales.

- max_p will not go underneath a packet loss probability of 1% and it will not exceed a packet loss probability of 50%. This is done to maintain acceptable performance even during a transient period (as the result of adapting $\max_p$ slowly) where the average queue length moves to the target zone.

- Whereas the original proposal in (Feng et al. 1999) varied $\max_p$ by multiplying it with constant factors α and β, it is now additively increased and multiplicatively decreased; this decision was made because it yielded the best behaviour in experiments.

The other RED parameters are set as follows (see Section 3.7 for a discussion of their impact):

w_q: This parameter controls the reactiveness of the average queue length to fluctuations of the instantaneous queue. Since the average queue length is recalculated whenever a packet arrives, the frequency of which directly depends on the link capacity (i.e. the higher the capacity of a link, the more the packets per second can traverse it), this means that the reactiveness of the average queue length also depends on the capacity. This effect is unwanted: w_q should generally be tuned to keep RED reactiveness in the order of RTTs. In Adaptive RED, this parameter is therefore set as a function of the link capacity in a way that will eliminate this effect (more precisely, it is set to $1 - \exp(1/C)$ where C is the link capacity).

$\min_{th}$: This parameter should be set to *target delay* $* C/2$.

$\max_{th}$: This parameter is set to $3 * \min_{th}$, which will lead to a target average queue size of $2 * \min_{th}$.

It is specifically stated in (Floyd et al. 2001) that the goal was not to come up with a perfect AQM mechanism; rather, the authors wanted to show that the average queue length can be stabilized and the problem of setting parameters can be circumvented without totally diverging from the original design of RED. At the same time, simulation results indicate that Adaptive RED is beneficial and remains robust in a wide range of scenarios.

4.4.2 Dynamic-RED (DRED)

Dynamic-RED (DRED) is a mechanism that stabilizes the queue of routers; by maintaining the average queue length close to a fixed threshold, it manages to offer predictable performance while allowing transient traffic bursts without unnecessary packet drops. The design of DRED is described in (Aweya et al. 2001); it follows a strictly control-theoretic approach. The chosen controller monitors the queue length and calculates the packet drop probability using an integral control technique, which will always work against an error (this is the measured output of the system, which is affected by perturbations in the environment, minus the reference input) in a way that is proportional to the time integral of the error, thereby ensuring that the steady-state error becomes zero. The error signal that is used to drive the controller is filtered with an EWMA process, which has the same effect as filtering (averaging) the queue length – just like RED, this allows DRED to accommodate short traffic bursts.

DRED has quite a variety of parameters that can be tuned; on the basis of analyses and extensive simulations, recommendations for their default values are given in (Aweya et al. 2001). Among other things, this concerns the sampling interval, which should be set to a fraction of the buffer size and not as high as the link capacity permits in order to allow the buffer to absorb 'noise' (short traffic bursts). Like standard RED, it has the goal of informing senders of congestion early via a single packet drop instead of causing a long series of drops that will lead to a timeout; this is achieved by reacting to the average and not to the instantaneous queue length.

4.4.3 Stabilized RED (SRED)

Stabilized RED (SRED) also aims at stabilizing the queue length, but the approach is quite different from DRED: since the queue oscillations of RED are known to often depend on the number of flows, SRED estimates this number in order to eliminate this dependence. This is achieved without storing any per-flow information, and it works as follows: whenever a new packet arrives, it is compared with a randomly chosen one that was received before. If the two packets belong to the same flow, a 'hit' is declared, and the number of 'hits' is used to derive the estimate. Since the queue size should not limit the chance of noticing packets that belong together, this function is not achieved by choosing a random packet from the buffer – instead, a 'zombie list' is kept.

This works as follows: for every arriving packet, a flow identifier (the 'five-tuple' explained in Section 5.3.1) is added to the list together with a timestamp (the packet arrival time) and a 'Count' that is initially set to zero. This goes on until the list is full; then, the flow identifier of arriving packets is compared to the identifier of a randomly picked entry in the list (a so-called 'zombie'). In case of a 'hit', the 'Count' of the zombie is increased by one – otherwise, the zombie is overwritten with the flow identifier of the newly arrived packet with probability p.

SRED was proposed in (Ott et al. 1999), where the timestamp is described as a basis for future work: in case of a non-hit, the probability of overwriting zombies could be made to depend on the timestamp, for example, older ones could be overwritten with a higher probability. This was, however, not included in the simulations that are reported in this paper.

The number of flows N is estimated with an EWMA process that takes a function 'Hit(t)' as its input, which is 1 in case of a hit and 0 otherwise; The weighting factor in this calculation (the same as α in Equation 3.1) depends on p above and the size of the zombie list. The drop probability is then calculated from the instantaneous queue length (the authors of (Ott et al. 1999) did not see a performance improvement of their scheme with the average queue length and state that it would be a simple extension) and N; assuming that only TCP is used and on the basis of some assumptions about the behaviour of this protocol, it is derived that for a certain limited range the drop probability must be of the order of N^2.

The final rule is to first calculate a preliminary dropping probability p_{sred}, which is set to one of the following: (i) a maximum (0.15 by default) if the current queue length is greater or equal to one-third of the total buffer size, (ii) a quarter of this maximum if it is smaller than a third but at least a sixth of the buffer size, or (iii) zero if it is even smaller. This appropriately limits the applicable probability range for incorporating the number of flows into the calculation. Then, the final drop probability is given by p_{sred} scaled with a constant and multiplied with N^2 if the number of active flows is small; otherwise, p_{sred} is used as it is.

The 'hit' mechanism in SRED has the additional advantage that it can be used to detect misbehaving flows, which have a higher probability of yielding a 'hit' than standard TCP flows do. This can simply be detected by searching the zombie list for entries with a high 'Count', and it could be used as a basis for protecting responsive flows from unresponsive ones.

4.4.4 BLUE

BLUE was the first AQM mechanism that did *not* incorporate the queue length in its packet loss probability calculation according to (Feng et al. 2002b), which also explains that, as a well-known fact from queuing theory, the queue length only directly relates to the number of active sources – and hence the actual level of congestion – when packet interarrivals have a Poisson distribution.[9] This is, however, not the case in the Internet (see Section 5.1 for further details), and so the scheme relies on the history of packet loss events and link utilization in order to calculate its drop probability. If the buffer overflows, the marking probability is increased, and it is decreased when the link is idle. More precisely, whenever such a 'loss' or 'link idle' event occurs and more than *freeze_time* seconds have passed, the drop probability is increased by δ_1 or decreased by δ_2, respectively. The authors of (Feng et al. 2002b) state that the parameter *freeze_time* should ideally be randomized in order to eliminate traffic phase effects but was set to a fixed value for their experiments; δ_1 was set to a significantly larger value than δ_2 to make the mechanism quickly react to a substantial increase in traffic load.

On the basis of BLUE, another mechanism called *Stochastic Fair Blue (SFB)* is also described in (Feng et al. 2002b). The goal of SFB is to protect TCP from the adverse influence of unresponsive flows by providing fairness among them, much like *Stochastic Fair Queuing (SFQ)*, a variant of 'Fair Queuing' (see Section 5.3.1) that achieves fairness – and therefore protection – by applying a hash function. However, whereas SFQ uses the hash function to map flows into separate queues, SFB maps flows into one out of N bins that are merely used to keep track of queue-occupancy statistics. In addition, there are L levels, each of which uses its own independent hash function; packets are mapped into one bin per level. The packet loss probability is calculated as with regular BLUE, but for each bin (assuming a certain fixed bin size). If a flow is unresponsive, it will quickly drive the packet loss probability of every bin it is hashed into to 1, and similarly, a responsive flow is likely to be hashed into at least one bin that is not shared with an unresponsive one. The decision of dropping a packet is based upon the minimum packet loss probability of all the bins that a flow is mapped into, and this will lead to an effective 'punishment' (a much higher drop probability) of unresponsive flows only.

4.4.5 Adaptive Virtual Queue (AVQ)

The *Adaptive Virtual Queue (AVQ)* scheme, presented in (Kunniyur and Srikant 2001), differs from the other mechanisms that we have discussed so far in that it does not explicitly calculate a marking probability; it maintains a virtual queue whose link capacity is less than the actual link capacity and whose buffer size is equal to the buffer size of the real queue. Whenever a packet arrives, it is (fictionally) enqueued in the virtual queue if there is space

---

[9]The relationship between the number of flows and the queue length is a common theme in AQM schemes.

available; otherwise, the packet is (*really*) dropped. The capacity of the virtual queue is updated at each packet arrival such that the behaviour of the algorithm is more aggressive when the link utilization exceeds the desired utilization and vice versa. This is done by monitoring the arrival rate of packets and not the queue length, which can therefore have the mechanism react earlier (before a queue even grows); the argument is the same as in Section 4.6.4, where it is explained why a congestion control mechanism that uses explicit rate measurements will typically outperform mechanisms that rely on implicit end-to-end feedback. Moreover, the reasons given in the previous section for avoiding reliance on the queue length also apply here.

The implementation of AVQ is quite simple: packets are not *actually* enqueued in the virtual queue – rather, its capacity (a variable) is updated on the basis of packet arrivals. This is quite similar to the 'token bucket' that is described in Section 5.3.1. There are only two parameters that must be adjusted: the desired utilization, which can be set using simple rules that are given in (Kunniyur and Srikant 2001), and a damping factor that controls how quickly the mechanism reacts – but, as pointed out in (Katabi and Blake 2002), properly setting the latter parameter can be quite tricky.

4.4.6 RED with Preferential Dropping (RED-PD)

Another approach to protect responsive flows from unresponsive ones is to actually store per-flow state, but only for flows that have a high bandwidth (i.e. flows that may be candidates for inferior treatment). In (Mahajan et al. 2001), this method is called *partial flow state* and applied in the context of an AQM mechanism that is an incremental enhancement of RED, *RED with Preferential Dropping (RED-PD)*. This scheme picks high-bandwidth flows from the history of RED packet drops, which means that it only considers flows that were already sent a congestion notification. Moreover, because of the removal of traffic phase effect from randomization, it can be assumed that flows are reasonably distributed in such a sample. Flows are monitored if they send above a configured target bandwidth; as long as the average queue length is above min_{th}, RED-PD drops packets from these flows before they enter the queue using a probability that will reduce the rate to the target bandwidth; The reason for doing so is that 'pushing down' flows with an unusually high bandwidth will allow others to raise theirs, thus equalizing the bandwidth of flows and making the mechanism one of many schemes that enforce fairness to at least some degree. The process is stopped when the average bandwidth is below the minimum threshold in order to always efficiently use the link.

Since the goal is to enforce fairness towards TCP, the target rate of RED-PD is set to the bandwidth that is obtained by a reference TCP flow. This is calculated with Equation 3.6; it was chosen because it is closer to the sending rate of a TCP flow (with no timeouts) over the short term than Equation 3.7, which may yield an estimate that is too low. After some derivations, the authors of (Mahajan et al. 2001) arrive at a rule to identify a flow by checking for a minimum number of losses that are spread out over a number of time intervals. If the dropping probability of a flow is not high enough for it to have its rate reduced to less than the target bandwidth with RED-PD, it is increased by the mechanism; if, on the other hand, the flow reduces its rate and did not experience a RED drop event in a number of time intervals, its drop probability is decreased. This ensures that the drop probability converges to the right value for every monitored flow.

4.4.7 Flow Random Early Drop (FRED)

If it is acceptable to maintain per-flow state because the number of flows is bounded and a large amount of memory is available, fairness can be enforced by monitoring all individual flows in the queue and the result can be used to make appropriate decisions. This is the approach taken by *Flow Random Early Drop (FRED)* (Lin and Morris 1997): this mechanism, which is another incremental RED enhancement, always accepts flows that have less than a minimum threshold min_q packets buffered as long as the average queue size is smaller than max_{th}. As with standard RED, random dropping comes into play only when the average queue length is above min_{th}, but with FRED, it only affects flows that have more than min_q packets in the queue. Note that this type of check requires the mechanism to store per-flow state for only the flows that have packets in the queue and not for all flows that ever traversed the link, and thus, the required memory is bounded by the maximum queue length.

FRED 'punishes' misbehaving flow via the additional variable max_q, which represents the number of packets that a flow may have in the buffer. No flow is allowed to exceed max_q, and if it tries to do so (i.e. a packet arrives even though max_q packets are already enqueued), FRED drops the incoming packet and increases a per-flow variable called *strike*. This is how the mechanism detects just how unresponsive a flow is; if *strike* is large (greater than one in the pseudo-code that is given in (Lin and Morris 1997)), a flow is not allowed to enqueue more than the average per-flow queue length (the average queue length divided by the number of flows found in the queue).

There are some more subtleties in the algorithm; for instance, it changes the fact that RED does not take departure events into account when calculating the average queue length. That is, when a packet arrives and the queue is very long, and then no packets arrive for a long time (allowing it to drain), the calculation upon arrival of the next packet will be based on the old (long) average queue length plus the instantaneous (short) one, leading to an unnecessarily high result. With FRED, the averaging also takes place whenever a packet leaves the queue and thus the result will be much lower in such a scenario.

4.4.8 CHOKe

We now turn to a mechanism that is designed to solve the fairness problem as simply as possible even though it has quite a complex acronym: *CHOose and Keep for responsive flows, CHOose and Kill for unresponsive flows (CHOKe)* (Pan et al. 2000). It resembles SRED in that it also compares incoming packets with randomly chosen ones in the queue, but it is much simpler and easier to implement – the idea of the authors was to show that the contents of the queue already provide a 'sufficient statistic' for detecting misbehaving flows, and applying complex operations on these data is therefore unnecessary. In particular, the queue probably buffers more packets that belong to a misbehaving flow, and it is therefore more likely to randomly pick a packet from such a flow than a packet from a properly behaving one. This reasoning led to the following algorithm that is run whenever a packet arrives:

- If the average queue length (as calculated with RED) is below min_{th}, the packet is admitted and the algorithm ends.

- Otherwise, a random packet is picked from the queue and compared with the newly arrived one. If the packets belong to the same flow, they are both dropped.

- Otherwise, the algorithm proceeds like normal RED: if the average queue size is greater than or equals max_{th}, the packet is dropped, and otherwise, it is dropped with a probability that is computed as in RED.

According to (Pan et al. 2000), it may also make sense to compare the incoming packet with not only one, but a number of packets from the buffer and drop all the ones that belong to the same flow; this imposes a more severe form of 'punishment' on flows that are unresponsive but requires more computation effort.

4.4.9 Random Early Marking (REM)

Random Early Marking (REM) is quite a well-known mechanism that was documented in a number of publications; the following description is based upon (Athuraliya et al. 2001). The first idea of REM is to stabilize the input rate around the link capacity and keep the queue small, no matter how many flows there are. The mechanism maintains a variable called *price*, which can be regarded as a model for the price that users should be charged for using a link: in times of congestion, the price should be high. The price variable is updated on the basis of the rate mismatch (the aggregate input rate minus the available capacity) and queue mismatch (the current queue length minus the target queue length), respectively. Depending on a weighted sum of these two values, the price is increased or decreased; as is common in AQM schemes, parameters that can be used to fine-tune the control (responsiveness and utilization versus queuing delay) as well as its previous value are also included in the update of this variable. For the price to stabilize, the weighted sum must be zero, that is, the queue length must exactly equal its target length and the input rate must be the same as the link capacity.

Ideally, a queue should always be empty, so zero seems to be a reasonable target queue length, and REM indeed supports setting it so. However, if the target queue length is greater than zero, it is also possible to refrain from monitoring the rate (because this can be a troublesome operation) and only use the queue length as an input factor for driving the control.[10] Even then, it is different from RED because it decouples queue control from measuring congestion – RED requires the average queue length to rise in order to determine significant congestion, but REM manages the same while keeping the queue stable. This last point is important, so here is another way of explaining it: the congestion measure that is embodied in the probability function of RED, the queue length, automatically grows as traffic increases. Averaging (filtering out short-term fluctuations) cannot change the fact that it is not possible to stabilize the queue at a low value while noticing that the amount of traffic significantly rises with RED. This is different with REM, which uses the (instantaneous) queue length only to explicitly *update* the congestion measure (price), but does not need to maintain it at a high value in order to detect significant traffic growth.

The drop probability is calculated as $1 - \phi^{-p(t)}$, where $p(t)$ is the price at time t – while RED uses a piecewise linear function, the drop probability of REM rises in

---

[10]This variant of REM is equivalent to the *PI (proportional-plus-integral)* controller described in (Hollot et al. 2001).

an exponential manner. This is required for the end-to-end drop probability to rise in the sum of the link prices of all the congested links along the path; in REM, it is actually equal to the drop probability function given above, but with $p(t)$ being the *sum* of all the drop probabilities per link, which is approximately proportional to the sum of link prices in the path. This means that REM implicitly conveys this aggregate price to end users via its packet dropping probability, which is another interesting feature of the mechanism.

4.4.10 Concluding remarks about AQM

As we have seen in the previous sections, there is an extremely diverse range of proposals for active queue management. They were presented in a relatively unordered manner, and there is a reason for this: it would of course be possible to categorize them somehow, but this is quite difficult as the classes that mechanisms would fit in are hardly totally disjunct. There are always schemes that satisfy several criteria at the same time. For example, if one was to separate them by their main goal (stabilizing the queue length at a low value versus providing fairness), then this would be an easy task with CHOKe and DRED, but difficult with SFB, which really is a mixture that tackles both problems. Classifying mechanisms on the basis of their nature is also not easy – some of them build upon RED while others follow an entirely different approach, and there is a smooth transition from one extreme end to the other. Also, some schemes base their decisions on the queue length while others measure drop or arrival rates. A rough overview of the AQM mechanisms is given in Table 4.1.

We have also only covered AQM schemes that drop (or perhaps ECN mark) packets if it is decided to do so, and if a distinction between flows is made, this only concerns unresponsive versus responsive ones. This is not the end of the story: as we will see in Chapter 5, there is a significant demand for classifying users according to their 'importance' (this is typically related to the amount of money they pay) and grant different service levels; whether a packet is 'important' or not is detected by looking at the IP header. Active queue management can facilitate this discrimination: there are variants of RED such as *Weighted RED (WRED)*, which maintains different values of max_p for different types of packets, or *RED with In/Out (RIO)*, which calculates two average queue lengths – one for packets that are 'in profile' and another one for packets that are 'out of profile' (Armitage 2000).

Clearly, the mechanisms described in this section are just a small subset of the AQM schemes found in the literature; for instance, AVQ is not the only one that maintains a virtual queue (a notable predecessor is described in (Gibbens and Kelly 1999)), and while the schemes in this section are among the more-popular ones and they are a couple of years old, research has by no means stopped since then. Examples of more-recent efforts can be found in (Heying et al. 2003), (Wang et al. 2004) and (Paganini et al. 2003), which describes an integrated approach that tackles problems of queue management and source behaviour at the same time. Also, just like TCP, AQM mechanisms have been analysed quite thoroughly; an example that provides more insight into the surprisingly good results attained with CHOKe is (Tang et al. 2004). Another overview that is somewhat complementary to the one found here is given in (Hassan and Jain 2004).

Table 4.1   Active queue management schemes

Mechanism	What is monitored?	What is done?
RED, Adaptive RED, DRED	Queue length	Packets are randomly dropped based upon the average queue length
SRED	Queue length, packet header	Flow identifiers are compared with random packets in a queue history ('zombie list'); this is used to estimate the number of flows, which is an input for the drop function
BLUE	Packet loss, 'link idle' events	The drop probability is increased upon packet loss and decreased when the link is idle
SFB	Packet loss, 'link idle' events, packet header	Packets are hashed into bins, BLUE is applied per bin, the minimum loss probabilities of all bins that a packet is hashed into is taken; it is assumed to be very high for unresponsive flows only
AVQ	Packet arrival rate	A virtual queue is maintained; its capacity is updated on the basis of packet arrival rate
RED-PD	Queue length, packet header, packet drops	The history of packet drops is checked for flows with a high rate; such flows are monitored and specially controlled
FRED	Queue length, packet header	Flows that have packets buffered are monitored and controlled using per-flow thresholds
CHOKe	Queue, packet header	If a packet is from the same flow as a randomly picked one in the queue, both are dropped
REM	Arrival rate (optional), queue length	A 'price' variable is calculated on the basis of rate (too low?) and queue (too high?) mismatch; the drop probability is exponential in price, the end-to-end drop probability is exponential in the sum of all link prices

4.5 Congestion control for multimedia applications

TCP provides reliable congestion-controlled data delivery. While it should be clear by now that congestion control is important for keeping the network stable (see the beginning of Chapter 2), reliability is a different issue. Most applications that we nowadays use in the Internet – web browsers, ftp downloads, email, instant messengers and so on – require reliable data transmission. If a packet is dropped in the network, TCP retransmits it, which will add some delay but is nevertheless highly desirable as all these applications would otherwise have to take care of retransmissions themselves. There are, however, applications where this function is a hindrance – thus, they are usually built on top of UDP. Take a VoIP tool, for example; when two people talk on the phone, having a delay in the communication of speech is highly undesirable for them (note that retransmissions in TCP cause more than just the single retransmitted packet to be delayed – this is the head-of-line blocking delay problem, which is explained in Section 3.6).

On the other hand, packet drops deteriorate speech quality somewhat because parts of the audio information are missing. This is less of a problem for VoIP; in fact, humans have built-in selective reliability that is superior to what a network protocol can provide – they use semantic interpretation to detect when to request retransmission. That is, we typically say 'I beg your pardon?' only when speech quality reaches a level where important information is lost. If, say, a single 'the' is missing from a sentence, the listener will normally not complain because the relevant information is still conveyed. Interacting humans (or at least humans on the telephone) therefore can be considered to have a control loop of their own that does not call for replication underneath.

So what about non-interactive multimedia applications, then? Should a one-way radio stream be reliably or unreliably transferred over the Internet? From a network viewpoint, there are some important differences between such one-way applications and interactive ones as well as live media versus playing prerecorded data. Let us examine four different scenarios where user A wants to say something to user B, and the audio data are transferred over the Internet with increasingly stringent network requirements:

1. *User A records what she has to say into a file, which is then transferred. The file is played only when it has completely arrived at the receiver.*
   This is the easiest case; it is a regular file download, and there are no strict delay or bandwidth requirements, that is, it is assumed that user B will simply wait for as long as it takes for the file to arrive.

2. *User A records what she has to say into a file, which is then transferred. Parts of the file are played as they arrive.*
   Even if packets are sent into the Internet at a constant rate, this is not necessarily how they reach the other end. There may be short term interruptions because packets are enqueued only to cause a burst when they leave the queue later. Since it is usually acceptable for user B in this scenario to wait for a while at the very beginning of the process (the connection setup phase), it makes sense to compensate for such problems via a buffer. Data can be played from this buffer at a regular rate while it is filled in a fluctuating manner; such a buffer effectively smoothes the traffic. Note that the playout rate must match the average throughput in this scenario: if the buffer is drained too fast, it can become empty (which means that the audio has to

be interrupted), and if the packet arrival rate is too high, the buffer overflows. Since throughput depends on the sending rate, this already means that some feedback to the sender, for example, via RTCP (Group et al. 1996), is necessary even if we neglect the requirement for network congestion control. Also note that the size of the buffer must match the timescale of fluctuations while it should not be too large at the same time – for ideal operation, the buffer should not be empty at the beginning, and thus a large buffer goes with severe initial delay during the connection setup phase.

3. *As user A speaks into the microphone, the audio data are transferred over the network to user B, where the data are played out as they arrive.*
All the requirements of scenario two are also there in this scenario, but the data rate of the sender underlies an additional restriction here. Multimedia data often have inherent rate fluctuations of their own, which depend on the actual content (e.g. MPEG mainly saves space by storing differences between video frames – a lot of movement means more data). While the sender rate in scenario 2 can be dictated by the network, how much data can be transferred also depends on what actually happens (e.g. if user A takes a break or keeps on talking) in this scenario.

4. *Users A and B have a telephone conversation over the Internet.*
Reliable transmission *might* make sense in scenarios 2 and 3 in order to enhance the quality if time permits (i.e. the retransmitted packet arrives at the receiver before it would have to be taken from the buffer), but efficiently implementing this can greatly complicate things. As explained above, however, retransmissions are typically out of the question in this scenario because delay is much more disturbing than packet loss.

TCP is all right for scenario 1, and it could also be possible to use it for scenarios 2 and 3 although it does not seem to be a very good match here because its strictly reliable operation is rather a hindrance. For scenario 4, the choice of TCP is really a poor one because of the potential delay incurred by (quite unnecessary) retransmissions. Thus, such applications typically use UDP, which does not provide congestion control. This means that programmers of such applications have to write the necessary functions themselves, which is quite a burden. There are several approaches to tackle this problem, and they are discussed in Sections 4.2.2, 4.5.2, 4.6.5 and 6.3; an overview is given by Figure 6.6. For now, we simplify by assuming that an application programmer indeed decides to implement congestion control functions on top of UDP and focus on the question, 'what should such a mechanism look like?'

When we mix the requirements of the scenarios above with the requirement of congestion control, two major problems become evident:

1. *Scenario 2:* The buffer at the receiver can only compensate for traffic fluctuations of a given interval; because of the self-similar nature of Internet traffic, it can be very hard, if not impossible, to come up with a guaranteed ideal buffer length (see Section 5.1).

2. *Scenario 3:* The source may not always be 'greedy', that is, fully utilize the allowed rate. When this happens, it is not possible to precisely sample the network conditions and cautious behaviour is recommendable (e.g. decaying the congestion window as explained for TCP in Section 4.1.3). Then again, the source may want to send at a high

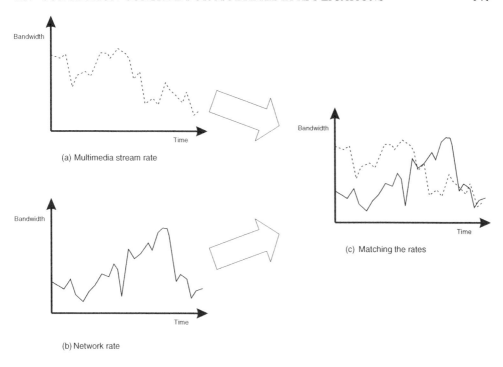

Figure 4.5 Matching a fluctuating application stream onto a fluctuating congestion control mechanism

rate when the allowed rate of the network is small, which means that data must either be dropped or buffered at the sender side. This problem is shown in Figure 4.5: (a) depicts an (artificial) fluctuating rate of a media stream, and (b) depicts an (artificial) fluctuating allowed sending rate of a congestion control mechanism. When these rates are combined as in (c), this can lead to a mismatch – the application may want to send at a high rate when only a low rate is allowed and vice versa.

The strict delay requirement of scenario 4 imposes upper limits on the feasible buffer lengths and therefore aggravates the situation.

There are two basic approaches to solve these two problems: first, it is possible to enforce a constant bit rate behaviour at the application level – this can, for instance, be done with the H.264 video codec – albeit this is usually not without cost (e.g. it is just a fact of life that video frames often do not change much in, say, a video conferencing setting, and it would seem stupid not to exploit this fact). Also, as we will see in the following sections, it is possible to use a congestion control scheme that fluctuates somewhat less than TCP. The ideal situation is shown in Figure 4.6, where a constant bit rate application stream is matched onto a constant bit rate congestion control mechanism that allows a larger rate. Clearly, such a scenario is impossible – not reacting at all is at odds with the very idea of congestion control – but one can approach this goal by trying to use somewhat *smoother* application streams and congestion control mechanisms.

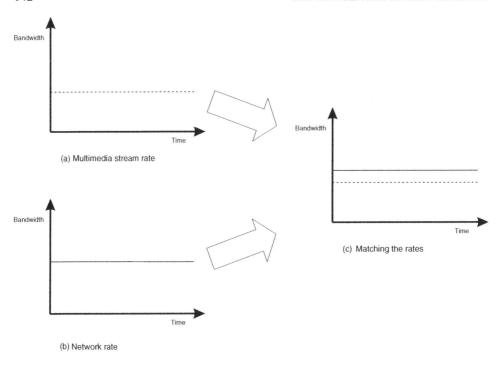

Figure 4.6   Matching a constant application stream onto a constant congestion control mechanism

Figure 4.7 shows a more-realistic scenario: the rate of the media stream is *adapted* to the rate of the congestion control mechanism. Such an application is accordingly called *adaptive*. This can be beneficial: sending at a high rate would correspond with a high packet loss ratio, which may yield inferior quality than using a lower rate in the long run. The rate could be changed by altering the compression factor or using a layered encoding scheme as described in Section 2.15. Simply put, users may prefer an uninterrupted video stream at a low quality over a high-quality stream with a lot of interruptions (a very low frame rate). On the other hand, quality fluctuations are also very undesirable from the user perspective (Mullin et al. 2001), and so it turns out that building a good adaptive multimedia application is a very tricky task. For instance, the buffering issues described above become more complicated when the application is adaptive – how to ideally handle a buffer when the underlying congestion control mechanism fluctuates is explained in (Rejaie et al. 1999a). Adaptive multimedia applications are a very interesting research topic that is right on the borderline between multimedia and computer networks; so, at this point, we turn to congestion control again.

From the discussion above, we can see that a smooth sending rate of a congestion control mechanism is desirable because it facilitates application/network rate matching operations; this is also one of the findings in (Feamster et al. 2001). On the other hand, it seems to be obvious that there is a trade-off between smoothness and reactiveness – the smoother a congestion control mechanism, the slower its reaction to changing network conditions – and indeed, this is shown via simulations in (Yang et al. 2001). So, depending on the type of

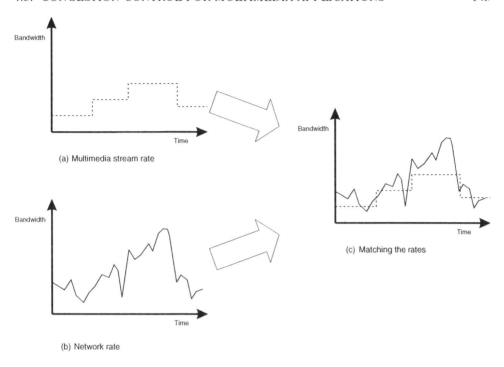

Figure 4.7 Matching an adaptive application stream onto a fluctuating congestion control mechanism

application that is used, it would be good to allow some flexibility, that is, decide just how reactive or smooth a mechanism should be. Additionally, a mechanism should be TCP-friendly if it is to be used in the public Internet without causing harm to a large number of TCP flows (see Section 2.17.4). Note that the definition for a flow to be TCP-friendly does not say that it must be so at every time point – 'not exceeding the rate of a conforming TCP' is assumed to be averaged over time intervals of several seconds or even the entire life of a flow.

In what follows, we will look at some mechanisms that were mainly designed to follow the goals of smoothness and TCP-friendliness as well as an effort to provide more flexibility. Recall from Section 4.2.2 that all these considerations do not solve the problem of placing additional burden onto programmers, and realizing congestion control on top of UDP is not ideal. We will therefore discuss a proposed solution that is currently being standardized in the IETF in Section 4.5.2. Finally, multimedia applications sometimes have a one-to-many communication model – think of a live TV broadcast across the Internet, for example; we will conclude this section with a brief look at some congestion control mechanisms for such scenarios.

4.5.1 TCP-friendly congestion control mechanisms

It seems to me that the research topic of unicast TCP-friendly congestion control for multimedia applications had a boom a couple of years ago, which recently ceased. Some of the proposals from the early 2000s or even earlier are still well known, and they certainly deserve a brief explanation here. However, apparently, serious effort was put into one of them only in recent years – TFRC – and many researchers who originally tried to smoothen the congestion response now moved on to different things, such as enhancing congestion control across high-speed links (Section 4.6) or updating their previous work to make them applicable for multicast scenarios (Section 4.5.3). I therefore decided to present only three somewhat representative 'historic' mechanisms before discussing an interesting undertaking to achieve more flexibility and then explaining TFRC; a thorough overview of TCP-friendly for unicast and multicast scenarios can be found in (Widmer et al. 2001). Here is something worth keeping in mind before we delve into the details: all of these mechanisms are pure end-to-end schemes, that is, they solely rely on explicit feedback that comes from a receiver that ACKs packets.

The Rate Adaptation Protocol (RAP)

First and foremost, what we want from a TCP-friendly congestion control mechanism is that it (i) is fair towards TCP and (ii) does not retransmit dropped packets. It may seem that the easiest method to achieve this is to start with TCP itself and remove its reliability, that is, retain the congestion control behaviour yet always use new data instead of retransmitting packets. This may sound easier than it is – recall the idea of reducing the rate in response to three DupACKs, mechanisms such as NewReno from the previous chapter, which waits for an ACK with a special sequence number, or spurious timeout detection, which is also based on retransmissions. All these mechanisms would need to be updated. While such endeavours were really undertaken (Jacobs and Eleftheriadis 1997), they inevitably inherit the bursty nature of the window-based flow control mechanism in TCP, which may not be very suitable for a media stream – in scenarios 3 and 4 in the previous section, the data cannot be stopped from being generated when the sender window is empty.

The *Rate Adaptation Protocol* (RAP) solves this problem by mimicking the AIMD behaviour of TCP in a rate-based fashion (Rejaie et al. 1999b). A receiver ACKs every packet. On the basis of this feedback, the sender explicitly calculates and controls the time between sending two packets, which is called the *inter-packet-gap* (IPG) – this variable is decreased every RTT (RAP estimates the RTT like TCP) using a 'step height' α, and it is divided by β (typically 0.5) in response to loss. Packet loss is detected via an 'unreliable equivalent' of three DupACKs: an ACK that implies reception of three packets after one that is missing. In addition, there is a timeout mechanism as in TCP.

RAP is a relatively simple scheme. It is one of the first of its kind – its authors were only concerned with TCP-friendliness and rate-based control, and in (Rejaie et al. 1999a), they showed that it can be embedded in an end-to-end architecture for real-time adaptive multimedia data transmission with intelligent buffer control. Smoothness was not yet a major issue.

The Loss-Delay – based Adjustment Algorithm (LDA+)

The Loss-Delay – based Adjustment Algorithm (LDA+) is also a rate-based mechanism that exhibits AIMD behaviour, but it adjusts its increase and decrease factors[11] on the fly.

---

[11]This is α and β in the case of RAP – in fact, using these variable names is common in documents describing TCP-friendly congestion control mechanisms, and we will therefore stick with them from now on.

The goal is to attain a smoother rate than TCP while being TCP-friendly and avoiding loss (Sisalem and Wolisz 2000a).

LDA+ achieves this in a somewhat unusual manner. First, it uses RTCP for feedback. This is a major difference between this protocol and most of its competitors because of a rule in RFC 1889 (Group et al. 1996) that requires these messages to be scaled with the data rate such that they do not exceed a certain percentage (normally 5%) of payload traffic, with at least 5 s between sending two RTCP packets. This means that there is usually much less feedback than with a mechanism that has the receiver acknowledge every packet. Second, it applies the 'packet pair' approach (see Section 4.6.3) to obtain the capacity of the bottleneck link and incorporates this measure in its rate-update calculation. Additionally, the theoretical rate of a TCP flow is calculated using Equation 3.7 and used to appropriately limit the rate of the sender.

TCP Emulation at Receivers (TEAR)

In (Rhee et al. 2000), a mechanism called *TCP Emulation at Receivers (TEAR)* is described. Once again, the main goal is to smoothen the rate while being TCP-friendly. In TEAR, the receiver itself calculates the rate instead of informing the sender about the congestion signals. It is built upon an assumption that the authors call 'rate independence': a 'loss event' in TCP (which does not include multiple losses from a single window) does not depend on the sending rate. This may seem to be at odds with the fact that losses are often correlated in normal FIFO queues, but such events typically occur on a shorter timescale than an RTT, according to (Rhee et al. 2000). This is not proven in the document, but it may be of minor importance because the mechanism just appears to work surprisingly well.

TEAR emulates the complete TCP sender behaviour at the receiver: it maintains the *cwnd* and *ssthresh* variables and calculates exactly what a sender would do, including its linear increase, fast recovery and timeout mechanisms. Theoretically, if TEAR were to calculate the rate of the sender as $cwnd/RTT$ and feed it back in an ACK of every packet, and the sender would simply transmit at this rate, the outcome would very closely resemble the behaviour of standard TCP. This is, however, not desired – the rate should be smoothed, and TEAR does this by calculating a weighted moving average over the last couple of 'epochs'. An 'epoch' equals one TCP 'sawtooth', which is determined by the time when the emulated TCP at the receiver switches from one state to another (e.g. from congestion avoidance to slow start). The result is then fed back to the sender, and since it is averaged, this feedback can be sent less often than with TCP.

Generalizing TCP-friendly congestion control

There is a series of papers that represent a generalization of TCP-friendly congestion control – that is, they gradually relaxed the rules for conforming behaviour. This work is based on analytical modelling of TCP. *General AIMD (GAIMD)* marks the beginning (Yang and Lam 2000a): the authors extended the derivation of Equation 3.7 in (Padhye et al. 1998) with a more general AIMD rule, where $\alpha > 0$ and $0 < \beta < 1$. As we have seen in Section 2.5.1, this suffices to attain fair and stable behaviour in the simplified model of (Chiu and Jain 1989). On the basis of the final equation for TCP behaviour with these parameters, the following simple relationship is derived for a flow to be TCP-friendly:

$$\alpha = \frac{4*(1-\beta^2)}{3} \tag{4.3}$$

This means that in order to remain fair towards TCP, one must consider a trade-off between responsiveness and smoothness – for example, for $\beta = 7/8$, α must be set to 0.31.

In (Bansal and Balakrishnan 2001), more general rules are presented. The idea is to describe the window update of an increase/decrease mechanism as follows:

$$\text{\textit{Increase}: } x(t+1) = x(t) + \alpha/x(t)^k; \alpha > 0 \tag{4.4}$$

$$\text{\textit{Decrease}: } x(t+1) = x(t) - \beta * x(t)^l; 0 < \beta < 1 \tag{4.5}$$

where $x(t)$ is the window at time t, α and β are the increase and decrease factors and k and l are constants. Note that this equation can yield the same results as Equation 2.1 if k and l are set to the right values: $k = 0, l = 1$ yields AIMD, $k = -1, l = 1$ yields MIMD, $k = -1, l = 0$ yields MIAD and $k = 0, l = 0$ yields AIAD. There are of course many more ways to set these parameters – all the increase/decrease mechanisms that can be constructed in this manner are called *binomial congestion control algorithms* in (Bansal and Balakrishnan 2001), which also contains rules for choosing a TCP-friendly subset of such mechanisms; one of them is that the sum of k and l must equal one. Two possible variants are *inverse increase/additive decrease* (IIAD), where $k = 1$ and $l = 0$ and the increase rule is therefore inversely proportional to the current window, and *SQRT* ($k = 1/2, l = 1/2$), which has an increase/decrease law that is inversely proportional and proportional to the square root of the current window, respectively. Note that these two algorithms are nonlinear, which means that their trajectories in vector diagrams as in Figure 2.4 are curves and not straight lines.

Finally, the *Choose Your Response Function (CYRF)* framework is a generalization that is even broader than the aforementioned rules (Sastry and Lam 2002b). Here, the window update function is described as

$$\text{\textit{Increase}: } x(t+1) = x(t) + x(t)/f(x(t)) \tag{4.6}$$

$$\text{\textit{Decrease}: } x(t+1) = x(t) - x(t) * g(x(t)) \tag{4.7}$$

where $f(x)$ and $g(x)$ are two monotonically non-decreasing functions with $f(x) > 0$ and $0 < g(x) <= 1$ for all $x >= 1$. Every application of the increase/decrease rules above brings the system closer to fairness; it is also possible to make this framework more general by dropping the requirement for $g(x)$ to be monotonic – then, there can be unfair decrease steps, but the system will nevertheless converge to fairness in the long run (from one congestion event to the next). This set of rules is called *1-CYRF*. In (Sastry and Lam 2002b), they are applied to design a mechanism named *LOG*; its behaviour is 'between' SQRT and IIAD. While LOG is TCP-friendly, this is not necessarily true for all the protocols that can be constructed with CYRF and 1-CYRF; in (Sastry and Lam 2002a), the necessary rules are described and another TCP-friendly mechanism named *SIGMOID* is developed. SIGMOID increases the window by one and decreases it using the function $c * x(t)/(1 + e^{-a*(x(t)-k)})$, which, on a side note, is the solution of Equation 4.12 (Welzl 2003) – this is an interesting relationship.

All these efforts are generalizations of each other: GAIMD with $\alpha = 1$ and $\beta = 0$ is TCP, whereas GAIMD is itself a special case of binomial congestion control ($k = 0, l = 1$), which can in turn be constructed using the CYRF (or at least 1-CYRF) framework. In total, this allows for quite a bit of flexibility when trying to be TCP-friendly – one can simply tune AIMD parameters as a trade-off between smoothness and responsiveness with GAIMD or use the framework given by binomial congestion control or even CYRF to choose an entirely different behaviour while not having to worry about its convergence properties.

Equation-based congestion control: TFRC

The TCP-friendly Rate Control (TFRC) protocol, presented in (Floyd et al. 2000a), is perhaps the most-elaborate attempt at congestion control that uses Equation 3.7. The underlying idea is simple: the protocol is rate based, the sender constantly updates its sending rate on the basis of the equation, and the necessary input parameters (RTO, which is derived from an RTT estimate, and the steady-state loss event rate p) come from receiver feedback. That is, the receiver calculates the loss event rate and sends it back to the sender, where these messages are additionally used for RTT estimation with an EWMA process, much like ACKs in TCP with the Timestamps option. The RTO value could theoretically be calculated as in TCP, but experiments have shown that approximating this calculation by simply setting RTO to four times the RTT is good enough.

Its loss event rate calculation at the receiver is perhaps the most critical part of TFRC; according to (Floyd et al. 2000a), it went through several discussions and design iterations, and finally, the decision was to use what the authors refer to as the 'average loss interval method'. This works as follows:

- As with TCP, data packets carry sequence numbers. At the receiver, the arrival of at least three packets with a sequence number that is higher than the sequence number of an expected packet indicates that the expected packet is lost. If the packet arrives late, this can be used to appropriately recalculate the loss interval afterwards and 'undo' the false assumption of loss to some degree; this resembles spurious timeout detection and recovery of TCP as described in Section 4.1.5, but at the receiver. The requirement of waiting for three packets beyond the one that experienced congestion is dropped with ECN, where a flag in the header can immediately indicate that a reaction is in order.

- The idea is to mimic not the behaviour of an outdated TCP Reno sender but rather a more-modern implementation like NewReno or SACK, which only reduces its rate once in case of several packet drops from a single window. Thus, the loss event rate that is used for calculating Equation 3.7 should not reflect more than one packet loss per RTT. The receiver therefore needs to filter out packet loss that occurs on a shorter timescale. It maps these measurements to a 'loss event' record, where a 'loss event' consists of at most one lost packet per RTT – and it needs some knowledge of the RTT to do this. This is achieved by requiring the sender to write its RTT estimate into data packets. If a packet is marked with ECN as having experienced congestion, the timestamp in the packet is used to figure out if the new and a previous congestion event are at least an RTT apart. If a packet is actually lost (as detected via the next three incoming packets), its 'arrival time' is interpolated.

- The number of packets that were received between two loss events is called a *loss interval*, and a weighted moving average is calculated over the history of loss intervals. The weights are determined with an equation that yields weights of 1, 1, 1, 1, 0.8, 0.6, 0.4 and 0.2 for $n = 8$, where n is the number of loss intervals that are taken into consideration. This parameter controls the speed at which TFRC reacts to different congestion levels; it should not be significantly greater than eight according to RFC 3448 (Handley et al. 2003), and it is described as a 'lower bound' in (Floyd et al. 2000a) – eight just appears to be a recommendable value. The 'loss event rate' p for the equation is the inverse of the weighted average loss interval. This value is finally transmitted back to the sender.

TFRC has some additional interesting features. One of them is oscillation avoidance: in environments with a relatively small number of flows and simple FIFO queues, even a single flow can have a severe impact on the RTT. The TCP response function defines a behaviour that is inversely proportional to the measured RTT – thus, if a single flow causes the RTT to grow, it will have to reduce its rate, which will decrease the RTT and allow the flow to increase its rate again, and so on. This can be countered by using a small weight in the EWMA process for RTT estimation – but then, it is possible for longer-term oscillations to occur as a result of overshooting the available bandwidth. This is solved in TFRC by adding some delay-based congestion avoidance (we will see more on this in Section 4.6) in the short term: when queues grow, the most-recent RTT sample will be greater than the average RTT, and therefore it may be prudent to slightly reduce the rate in this case. This is achieved by multiplying the rate with the average RTT divided by the current RTT sample. The impact of this factor is further diminished by not actually using the RTT but by using the square root of the RTT wherever an RTT is used in this calculation (including the EWMA process).

Another necessary feature is a slow-start mechanism; since TFRC is rate based and therefore not ACK-clocked, there is some danger that a simplistic slow-start implementation would overshoot the available bandwidth by even more than TCP. This is solved by having the receiver feed back the rate at which packets arrived during the last RTT to the sender and using this measure to limit the new rate to the minimum of twice the current actual rate and twice the rate from the receiver. Additional protocol elements include a history discounting mechanism that allows a sender to react more rapidly to decreasing loss and a feedback timer that has the sender halve its rate in the absence of feedback.

TFRC is perhaps the most-thoroughly investigated TCP-friendly congestion control mechanism with a smooth sending rate – its behaviour was studied, for instance, in (Deepak et al. 2001) and (Vojnovic and Le Boudec 2002). Like all of the smooth congestion control schemes, it is only useful for applications that actually require such behaviour, as smoothness always comes at the cost of responsiveness (Yang et al. 2001); in particular, it does not aggressively probe for bandwidth and only increases its rate by approximately 0.14 packets per RTT (0.22 packets in slow start). TFRC is the only such mechanism that was taken to the IETF – it is specified in RFC 3448 (Handley et al. 2003), which does not describe a complete protocol but describes only a way of implementing the mechanism on top of other protocols such as UDP or RTP. A more-specific RTP variant is in the works (Gharai 2004), and TFRC is also available in the DCCP protocol that will be described in the next section. A recent paper describes how to make it useful for applications that need to vary the packet size (Widmer et al. 2004), and discussions in the IETF DCCP working group have led to ongoing work that ensures suitability for VoIP (Floyd and Kohler 2005).

4.5.2   The Datagram Congestion Control Protocol (DCCP)

As already explained, interactive multimedia applications put timely delivery before reliability, and therefore normally use UDP instead of TCP. Since UDP provides no congestion control, it is up to the application programmer to implement a suitable mechanism that is TCP-friendly (see Section 4.5 for some examples). Unfortunately, this is not an easy task, and it is questionable whether it is worth the effort if we assume that the developer (the application programmer, or the company that pays her) is selfish. This assumption is probably correct, as it is in line with the goal of maximizing profit – in other words, investing money in the implementation of a TCP-friendly congestion control scheme is probably not a good decision unless it is connected to a financial benefit. This may be the reason why, as we will see in Chapter 6, most UDP-based applications do not seem to realize TCP-friendly congestion control nowadays.

The IETF seeks to counter this development with the Datagram Congestion Control Protocol (DCCP), which is supposed to be used by such applications in the future. The idea is to relieve application programmers: they should not have to worry about congestion control but leave it up to the underlying transport protocol instead. By putting congestion control where it belongs – in the operating system – the necessary functions can also be carried out with the precise timing that may be required. At the time of writing, the protocol is a work in progress – there are no RFCs, but there is a working group that wrote several Internet-drafts over the last couple of years. Now, it seems that the discussions have converged and that the main specification will soon turn into a set of standards track RFCs. At first sight, it may seem strange that this single protocol is specified in more than one document; the reason for this separation is that the main DCCP specification (Kohler et al. 2005) does *not* contain a congestion control mechanism. Instead, it is a framework for schemes like the ones in the previous section. DCCP does not mandate TCP-friendliness of congestion control mechanisms, but such behaviour can probably be expected from them, given that DCCP is developed in the IETF, which has the position that all mechanisms should be TCP-friendly.

Functional overview

Most of the experimental congestion control mechanisms that were described in this chapter are merely *schemes* and not *protocols*. A transport protocol encompasses many functions that are not necessarily congestion control specific – usually, there are headers with sequence numbers for packet identification, checksums provide data integrity and port numbers ensure that several applications can communicate even when they share the same IP addresses, to name but a few. DCCP embeds all this functionality and more in its core protocol. According to (Kohler et al. 2005), one way of looking at DCCP is as TCP minus byte-stream semantics and reliability, or as UDP plus congestion control, handshakes and acknowledgements; here are the most-important things that DCCP does in addition to these functions:

Connection setup and teardown: This functionality is included even though the protocol is unreliable; the main goal is to facilitate middlebox traversal – for example, firewalls that tend to throw away all UDP packets may more selectively allow some DCCP connections. When initiating a connection, a *cookie* is used to protect against SYN floods. This may require some explanation: the well-known 'TCP SYN' flood is a denial-of-service attack that exploits the fact that TCP maintains state whenever a new connection is initiated – and per-flow state is known not to scale well. A web server, for instance, must remember some things regarding the connection that a browser wants to initiate. Cookies let the server place the information inside the packet instead – the protocol specifies that the other peer must then send this information back in the next packet. On a side note, there is a way to do this in TCP without changing the protocol specification – the initial sequence numbers can be used to convey the required information.[12]

Acknowledgements: Packets are ACKed in order to provide a congestion control mechanism with all the feedback that it may need. Since DCCP is unreliable and it therefore never retransmits a packet, it was decided that sending cumulative ACKs would not be a reasonable design choice. Yet, with full ECN nonce functionality and totally unreliable behaviour, ACKing individual packets only would require the receiver to maintain state that could theoretically grow without bounds. Therefore, unlike TCP, DCCP has reliable ACKs – when the receiver is informed that an ACK has made it, it can remove the related state. There is another advantage from reliable ACKs: usually, a single lost ACK goes unnoticed in TCP, and this will not cause much harm – but it has the disadvantage that it disables the protocol from carrying out congestion control along the ACK path. In case of highly asymmetric connections, this can lead to adverse effects that can be very hard to overcome (see Section 4.7). Reliable ACKs eliminate this problem; they are realized somewhat similar to reliable ECE and CWR communication in TCP, that is, the receiver keeps ACKing until the sender ACKs the ACK, and the sender will keep ACKing ACKs until the receiver stops sending the same ACK. There is no need for ACKs of ACKs of ACKs, but there is still a minor issue – ACKs themselves require sequence numbers in order to enable this functionality. Controlling congestion along the ACK path is done in a rate-based manner via a feature that is called *ACK Ratio*. The goal is not to realize exact TCP-friendliness but just to avoid congestion collapse from utterly unresponsive behaviour.

Feature negotiation: In DCCP, *features* are simply variables,[13] and four *options* are used to negotiate them: 'Change L' ('L' stands for 'local'), 'Change R' ('R' stands for 'Remote'), 'Confirm L' and 'Confirm R'. As the names suggest, 'Change' is a request to change a feature and 'Confirm' is a way of saying 'affirmative' in response to a Change request. The option format of DCCP will be explained in the next section. One especially important feature is CCID, the Congestion Control ID – this means that this mechanism can, for instance, be used to negotiate which congestion control mechanism is chosen. Some features may allow a Change option to carry a series of values as a preference list; the semantics of this is something like 'I would like

---

[12]More details can be found at http://cr.yp.to/syncookies.html
[13]Note that this meaning of the word 'feature' prevails throughout Section 4.5.2.

CCID 1, but otherwise, please use CCID 2. If you cannot even give me CCID 2, please use CCID 3'. The receiver can answer such a request with a Confirm option that also carries a preference list – which means that it says something like 'I picked CCID 2 – my own preference list is 5, 2, 1'. The main DCCP specification (Kohler et al. 2005) explains exactly how such a process is to be carried out; it also ensures that lost feature negotiation messages are retransmitted, rendering this mechanism reliable.

Checksums: There is a standard checksum that can be restricted to the header in order to realize UDP-Lite functionality and an additional 'Data Checksum' option (which is a stronger CRC-32c checksum) that enables a DCCP endpoint to distinguish corruption-based loss from other loss events; see Section 4.1.7 for more details.

Full duplex communication: Applications such as audio or video telephones may require a bidirectional data stream; in such cases, having DCCP support such operation makes it possible to piggyback ACKs onto data packets. The fact that this also simplifies middlebox traversal (firewalls only need to detect and allow one connection) was another reason to make DCCP a full duplex protocol. Operationally, a DCCP connection is logically broken into two logically independent 'half-connections', which independently negotiate features. Since the CCID is just another feature, this means that the two flows can even be regulated with two different congestion control mechanisms.

Security: Making the protocol at least as safe against misbehaviour as a state-of-the-art TCP implementation was an important design goal; among other things, this concerns attacks from misbehaving receivers that would like to achieve more than their fair share. For example, DCCP always uses the nonce mechanism that was described in Section 4.1.4, which also eliminates the possibility to carry out an 'optimistic ACK' attack as described in Section 3.5. 'ACK division' attacks (explained in the same section) are inherently impossible because DCCP counts packets and not bytes. This is true for all regular data packet and ACK sequence numbers alike. In order to prevent hijacking attacks, DCCP checks sequence numbers for validity. In TCP, this is a natural property of window-based flow control – DCCP defines a Sequence Window to do the same, but unreliability makes this a somewhat difficult task.

While TCP retransmits packets (and thereby reuses sequence numbers) in the presence of loss, a long lasting loss event causes a DCCP sender to increase the sequence number that it places in packets further and further. Therefore, it can never be guaranteed that sequence numbers stay within a pre-defined range, and a synchronization mechanism is required. Specifically, upon reception of a packet that carries a sequence number that lies outside a so-called Sequence Window, a DCCP entity asks its peer whether the sequence number is really correct and provides it with its own sequence number, which must be acknowledged in the reply. This simple mechanism provides some protection against hijackers that cannot snoop.

Significantly better security could be obtained via encryption, but this would be a duplication of functionality that is already provided by IPSec, and it was therefore not included in the protocol. Finally, some protection against denial-of-service attacks is provided by the aforementioned cookie that is used when setting up a connection.

Mobility: The question whether to include mobility in DCCP was discussed at length in the DCCP working group; eventually, mobility and multihoming were not included in the main document, and the specification was postponed. In (Kohler 2004), a rudimentary mechanism that slightly diverges from the original DCCP design rationale of not using cryptography is described; it is disabled by default, and an endpoint that wants to use this mechanism must negotiate enabling the corresponding feature. The semantics of the scheme in (Kohler 2004) are simpler than mobility support in SCTP and resemble Mobile IP as specified in RFC 3344 (Perkins 2002). In addition to the feature that is required for negotiating usage of this function, the specification defines a 'Mobility ID' feature and a 'DCCP-Move' packet type that is used to inform the other end of the connection when a host has moved.

Clearly, there are many more things that an application could want from DCCP – among them are selective reliability and packet-level FEC. There is an obvious trade-off between lightweight operation and functionality richness; since the protocol should be broadly applicable for future applications of an unknown kind, it was decided to refrain from embedding mechanisms that could just as well be implemented on top of DCCP.

Packet format

Header overhead is minimized by defining only a very small generic header that contains indispensable information for all packets; any additional functionality requires adding the corresponding subheader, which is identified via a 'type' field in the generic header. Note that there is only one generic header and one subheader in a packet – this is a major difference between DCCP and SCTP, where multiple chunks can be included in order to realize certain functions. If options are used, they are appended after the extension header, and a packet ends with application data if there are any (as with TCP, ACKs do not necessarily contain data). There are two variants of the generic header, depending on the size of the desired sequence number space. Whether short sequence numbers are used or long sequence numbers are used is indicated by the state of a flag called X; both variants are shown in Figure 4.8. In what follows, we will neglect this distinction and simply refer to the 'generic header'. The meaning of its fields is as below:

Source Port/Destination Port (16 bit each): These fields are similar to the same fields of TCP; see Section 3.1.1 for further details.

Data Offset (8 bit): This is the offset from the beginning of the DCCP packet to the data area.

CCVal (4 bit): This field is reserved for CCID-specific information – that is, it can be used by a congestion control mechanism to encode some information in packets.

Checksum Coverage (CsCov) (4 bit): Like the 'Checksum Coverage' field in UDP-Lite (see RFC 3828 (Larzon et al. 2004)), this makes it possible to restrict the range of the checksum to the header, the whole packet, or a certain range in between.

Checksum (16 bit): The checksum in the general header is a standard Internet checksum.

Type (4 bit): This field is used to indicate the subsequent extension header.

Source Port			Destination Port	
Data Offset	CCVal	CsCov	Checksum	
Res	Type	X = 1	Reserved	Sequence Number (high bits)
Sequence Number (low bits)				

Source Port			Destination Port
Data Offset	CCVal	CsCov	Checksum
Res	Type	X = 0	Sequence Number (low bits)

Figure 4.8 The DCCP generic header (both variants)

Additionally, there is a three-bit 'Reserved (Res)' field next to the 'Type' field and an eight-bit 'Reserved' field next to the sequence number if 'X' is zero. In accordance with the robustness principle that already guided the design of TCP, this field must be set to zero when a packet is sent and it is ignored by the receivers.

The main specification defines nine different subheaders that are identified via the 'Type' field: 'DCCP-Request', 'DCCP-Response', 'DCCP-CloseReq', 'DCCP-Close' and 'DCCP-Reset' for connection handling, 'DCCP-Data' for regular data packets, 'DCCP-Ack' for ACKs, 'DCCP-DataAck' for packets that carry both application data and ACK information, and 'DCCP-Sync' as well as 'DCCP-SyncAck', which are used to resynchronize sequence numbers as described in the previous section after a severe loss event. Since the general goal of splitting headers in this manner was to be more flexible and save some space (e.g. packets that carry no ACK information need no ACK sequence number), the subheaders are quite short. For instance, a DCCP-Data subheader only contains options in a format similar to IP and TCP and padding bits, while a DCCP-DataAck merely adds an acknowledgement number.

Options and features

Options, as the name suggests, are optional. A packet can carry several options at the same time, or none at all. The following options are defined:

Padding: As in other protocols like TCP or IP, this is a single-byte option that has no meaning but is used to align the beginning of application data to a 32-bit address.

Mandatory: This option indicates that the following option is mandatory; if the receiver does not understand the next option, it is required to send an error message back to the sender using the 'DCCP-Reset' packet type.

Slow Receiver: This option is a simple means to carry out flow control – it does not carry any additional information, and it tells a sender that it should refrain from increasing its rate for at least one RTT.

Change L/R, Confirm L/R: These options are used for feature negotiation as explained in the previous section. Actually, this is not an entirely simple process, and the specification therefore includes pseudo-code of an algorithm that properly makes a decision.

Init Cookie: This was explained on Page 150.

NDP Count: Since sequence numbers increment with any packet that is sent, a receiver cannot use them to determine the amount of application data that was lost. This problem is solved via this option, which reports the length of each burst of non-data packets.

Timestamp, Timestamp Echo and Elapsed Time: These three options help a congestion control mechanism to carry out precise RTT measurements. 'Timestamp' and 'Timestamp Echo' work similar to the TCP Timestamps option described in Section 3.3.2 of the previous chapter; the Elapsed Time option informs the sender about the time between receiving a packet and sending the corresponding ACK. This is useful for congestion control mechanisms that send ACKs infrequently.

Data Checksum: See Section 4.1.7.

ACK Vector: These are actually two options – one representing a nonce of 1 and one representing a nonce of 0. The ACK Vector is used to convey a run-length encoded list of data packets that were received; it is encoded as a series of bytes, each of which consists of two bits for the state ('received', 'received ECN marked', 'not yet received' and a reserved value) and six bits for the run length. For consistency, the specification defines how these states can be changed.

Data Dropped: This option indicates that one or more packets did not correctly reach the application; much like the ACK Vector, its data are run-length encoded, but the encoding is slightly different. Interestingly, with this option, a DCCP receiver can inform the sender not only that a packet was dropped in the network but also that it was dropped because of a protocol error, a buffer overflow in the receiver, because the application is not listening (e.g. if it just closed the half-connection), or because the packet is corrupt. The latter notification requires the Data Checksum option to be used; it is also possible to utilize it for detecting corruption but, nevertheless, hand over the data to the application – such things can also be encoded in the Data Dropped option.

Most of the features of DCCP enable negotiation of whether to enable or disable support of an option. For example, ECN is enabled by default, but the 'ECN Incapable' feature allows turning it off; similarly, 'Check Data Checksum' lets an endpoint negotiate whether its peer will definitely check Data Checksum options. The 'Sequence Window' feature controls the width of the Sequence Window described in the previous section, where we have already discussed CCID and 'ACK Ratio'; the remaining features are probably self-explanatory, and their names are 'Allow Short Seqnos', 'Send Ack Vector', 'Send NDP

Count' and 'Minimum Checksum Coverage'. The specification leaves a broad range for CCID-specific features.

Using DCCP

The DCCP working group is currently working on a user guide for the protocol (Phelan 2004); the goal of this document is to explain how different kinds of applications can make use of DCCP for their own benefit. This encompasses the question of what CCID to use. There are currently two CCIDs specified, CCID 2, which is a TCP-like AIMD mechanism, and CCID 3, which is an implementation of TFRC (see Section 4.5.1), but there may be many more in the future. CCID specifications explain what conditions it is recommended for, describe their own options, features and packet as well as ACK format, and of course explain how the congestion control mechanism itself works. This includes a specification of the response to the Data Dropped and Slow Receiver options, when to generate ACKs and how to control their rate, how to detect sender quiescence and whether ACKs of ACKs are required.

In the current situation, the choice is not too difficult: CCID 2 probes more aggressively for the available bandwidth and may therefore be more appropriate for applications that do not mind when the rate fluctuates wildly, and CCID 3 is designed for applications that need a smooth rate. The user guide provides some explanations regarding the applicability of DCCP for streaming media and interactive game applications as well as considerations for VoIP. It assumes that senders can adapt their rate, for example, by switching between different encodings; how exactly this should be done is not explained. In the case of games, a point is made for using DCCP to offload application functionality into the operating system; for example, partial reliability may be required when messages have different importance. That is, losing a 'move to' message may not be a major problem, but a 'you are dead' message must typically be communicated in a reliable manner. While DCCP is unreliable, it already provides many of the features that are required to efficiently realize reliability (ACKs, the Timestamp options for RTT calculation, sequence numbers etc.), making it much easier to build this function on top of it than developing all the required functions from scratch (on top of UDP).

The main advantage of DCCP is certainly the fact that most congestion control considerations could be left up to the protocol; additional capabilities such as Path MTU Discovery, mobility and multihoming, partial checksumming, corruption detection with the Data Checksum option and ECN support with nonces additionally make it an attractive alternative to UDP. It remains to be seen whether applications such as streaming media, VoIP and interactive multiplayer games that traditionally use UDP will switch to DCCP in the future; so far, implementation efforts have been modest. There are several issues regarding actual deployment of DCCP – further considerations can be found in Section 6.2.3.

4.5.3 Multicast congestion control

Traditionally, multicast communication was associated with unreliable multimedia services, where, say, a live video stream is simultaneously transmitted to a large number of receivers. This is not the only type of application where multicast is suitable, though – reliable many-to-many communication is needed for multiplayer games, interactive distributed simulation and collaborative applications such as a shared whiteboard. Recently, the success of one-to-many applications such as peer-to-peer file sharing tools has boosted the relevance of reliable multicast, albeit in an overlay rather than IP-based group communication context (see Section 2.15). IP multicast faced significant deployment problems, which may be due to the fact that it requires non-negligible complexity in the involved routers; at the same time, it is not entirely clear whether enabling it yields an immediate financial gain for an ISP. According to (Manimaran and Mohapatra 2003), this is partly due to a chicken – egg problem: ISPs are waiting to see applications that demand multicast whereas users or application developers are waiting for wide deployment of multicast support. This situation, which bears some resemblance to the 'prisoner's dilemma' in game theory, appears to be a common deployment hindrance for Internet technology (see Page 212 for another example).

Some early multicast proposals did not incorporate proper congestion control; this is pointed out as being a severe mistake in RFC 2357 (Mankin et al. 1998) – in fact, multicast applications have the potential to do vast congestion-related damage. Accordingly, there is an immense number of proposals in this area, and they are very heterogeneous; in particular, this is true for layered schemes, which depend on the type of data that are transmitted. The most-important principles of multicast were briefly sketched in Section 2.15, and a thorough overview of the possibilities to categorize such mechanisms can be found in RFC 2887 (Handley et al. 2000a). Exhaustive coverage would go beyond the scope of this book – in keeping with the spirit of this chapter, we will only look at two single-rate schemes, where problems like ACK filtering (choosing the right representative) are solved, and conclude with an overview of congestion control in layered multicast, where this function is often regulated via group membership only and the sender usually does not even receive related feedback.

Notably, the IETF does some work in the area of reliable multicast; since the common belief is that a 'one size fits all' protocol cannot meet the requirements of all possible applications, the approach currently taken is a modular one, consisting of 'protocol cores' and 'building blocks'. RFC 3048 (Whetten et al. 2001) lays the foundation for this framework, and several RFCs specify congestion control mechanisms in the form of a particular building block.

TCP-friendly Multicast Congestion Control (TFMCC)

TCP-friendly Multicast Congestion Control (TFMCC) is an extension of TFRC for multicast scenarios; it can be classified as a single-rate scheme – that is, the sender uses only one rate for all receivers, and it was designed to support a very large number of them. In scenarios with a diverse range of link capacities and many receivers, finding the perfect rate is not an easy task, and it is not even entirely clear how a 'perfect' rate would be defined. In order to ensure TCP-friendliness at all times, the position taken for TFMCC is that flows from the sender to any receiver should not exceed the throughput of TCP. This can be achieved by transmitting at a TCP-friendly rate that is dictated by the feedback of the slowest receiver. Choosing the slowest receiver as a representative may cause problems for the whole scheme in the face of severely impaired links to some receivers – such effects should generally be countered by imposing a lower limit on the throughput that a receiver must be able to attain. When it is below the limit, its connection should be closed.

In TFRC, the receiver calculates the loss event rate and feeds it back to the sender, where it is used as a input for the rate calculation together with an RTT estimate; in TFMCC, this whole process is relocated to the receivers, which then send the final rate back to the sender. In order to ensure that the rate is always dictated by the slowest receiver, the sender will immediately reduce its rate in response to a feedback message that tells it to do so; since such messages would be useless and it is important to reduce the amount of unnecessary feedback, receivers normally send messages to the sender only when their calculated rate is less than the current sending rate. Only the receiver that is chosen as the representative (called *current limiting receiver (CLR)* in TFMCC) because it attains the lowest throughput is allowed to send feedback at any time – this additional feedback is necessary for the sender to increase its rate, as doing so in the absence of feedback can clearly endanger the stability of the network. The CLR can always change because the congestion state in the network changes or because the CLR leaves the multicast group; the latter case could lead to a sudden rate jump, and therefore the sender limits the increase factor to one packet per RTT.

Calculating the rate at receivers requires them to know the RTT, which is a tricky issue when there are no regular messages going back and forth. The only real RTT measurement that can be carried out stems from feedback messages which are answered by the sender. This is done by including a timestamp and receiver ID in the header of payload packets. The sender decides for a receiver ID using priority rules – for example, a receiver that was not able to adjust its RTT for a long time is favoured over a receiver that was recently chosen. These actual RTT measurements are rare, and so there must be some means to update the RTT in the meantime; this is done via one-way delay measurements, for which the receivers and the sender synchronize their clocks, and this is complemented with sender-side RTT measurements that are used to adjust the calculated rate when the sender reacts to a receiver report. Two more features further limit the amount of unnecessary feedback from receivers:

- Each receiver has a random timer, and time is divided into feedback rounds. Whenever a timer expires and causes a receiver to send feedback, the information is reflected back to all receivers by the sender. When a receiver sees feedback that makes it unnecessary to send its own, it cancels its timer. Such random timers, which were already mentioned in Section 2.15, are a common concept (Floyd et al. 1997). Interestingly, TFMCC uses a randomized value which is biased in favour of receivers with lower rates.

  The echoed feedback is used by receivers to cancel their own feedback timer if the reflected rate is not significantly larger (i.e. more than a pre-defined threshold) than their own calculated rate. This further reduces the chance of receivers sending back an unnecessarily high rate.

- When the sending rate is low and loss is high, it is possible for the above mechanism to malfunction because the reflected feedback messages can arrive too late to cancel the timers. This problem is solved in TFMCC by increasing the feedback in proportion to the time interval between data packets.

A more detailed description of TFMCC can be found in (Widmer and Handley 2001); its specification as a 'building block' in (Widmer and Handley 2004) is currently undergoing IETF standardization with the intended status 'Experimental'.

pgmcc

The *Pragmatic General Multicast (PGM)* protocol realizes reliable multicast data transport using negative acknowledgements (NAKs); it includes features such as feedback suppression with random timers, forward error correction and aggregation of NAKs in PGM-capable routers (so-called network elements (NEs)) (Gemmell et al. 2003). While its specification in RFC 3208 (Speakman et al. 2001) encompasses some means to aid a congestion control mechanism, it does not contain a complete description of what needs to be done – the detailed behaviour is left open for future specifications. This is where *pgmcc* comes into play (Rizzo 2000). While it was developed in the context of PGM (where it can seamlessly be integrated), this mechanism is also modular; for instance, there is no reason why pgmcc could not be used in an unreliable multicast scenario. In what follows, we will describe its usage in the context of PGM.

Other than TFMCC, pgmcc is window based. Receivers calculate the loss rate with an EWMA process, and send the result back to the sender in an option that is appended to NAKs. Additionally, the option contains the ID of the receiver and the largest sequence number that it has seen so far. The latter value is used to calculate the RTT at the sender, which is not a 'real-time' based RTT but is merely calculated in units of seconds for ease of computation and in order to avoid problems from timer granularity. Since the RTT is only used as a means to select the correct receiver in pgmcc, this difference does not matter; additionally, simulation results indicate that time-based measurements do not yield better behaviour.

pgmcc adds positive acknowledgements (ACKs) to PGM. Every data packet that is not a retransmission must be ACKed by a representative receiver, which is called the *acker*. The acker is selected by the sender via an identity field that pgmcc adds to PGM data packets, and the decision is based on Equation 3.6 (actually, a slightly simplified form thereof that is tailored to the needs of pgmcc). This is because pgmcc emulates the behaviour of TCP by opening a window whenever an ACK comes in and reducing it by half in response to three DupACKs. This window is, however, not the same one that is used for reliability and flow control – it is only a means to regulate the sending rate. ACK clocking is achieved via a token mechanism: sending a packet 'costs' a token, and for each incoming ACK, a token is added. The transmission must be stopped when the sender is out of tokens – this could be regarded as the equivalent of a TCP timeout, and it also causes pgmcc to enter a temporary mode that resembles slow start.

Congestion control for layered multicast

As described in Section 2.15, layered (multi-rate) congestion control schemes require the sender to encode the transmitted data in a way that enables a receiver to choose only certain parts, depending on its bottleneck bandwidth. These parts must be self-contained, that is, it must be possible for the receiver to make use of these data (e.g. play the audio stream or show the video) without having to wait for the remaining parts to arrive. This is obviously highly content dependent – an entirely different approach may be suitable for a video stream than for hierarchically encoded control information in a multiplayer game, and not all data can be organized in such a manner. A good overview of layered schemes for video data can be found in (Li and Liu 2003); the following discussion is also partly based on (Widmer 2003).

The first well-known layered multicast scheme is *Receiver-driven Layered Multicast (RLM)* (McCanne et al. 1996), where a sender transmits each layer in a separate multicast group and receivers periodically join the group that is associated with a higher layer so as to probe for the available bandwidth. Such a 'join-experiment' can repeatedly cause packet loss for receivers who share the same bottleneck – these receivers must synchronize their behaviour. The way that this is done in RLM leads to long convergence time, which is a function of the number of receivers and therefore imposes a scalability limit. Additionally, RLM does not necessarily result in a fair bandwidth distribution and is not TCP-friendly. These problems are tackled by the *Receiver-driven Layered Congestion Control (RLC)* protocol (Vicisano et al. 1998), which emulates the behaviour of TCP by appropriately choosing the sizes of layers and regulating the group joining and leaving actions of receivers. These actions are carried out in a synchronous fashion; this is attained via specially marked packets that indicate a 'synchronization point'. Since there is no need for coordination among receivers, this scheme can converge much faster than RLM.

Mimicking TCP may be a feasible method to realize TCP-friendliness, but it is undesirable for a streaming media application, as we have seen in Section 4.5; this is true for multicast, just as unicast. One mechanism that takes this problem into account is the *Multicast Enhanced Loss-Delay Based Adaptation Algorithm (MLDA)* (Sisalem and Wolisz 2000b), which, as the name suggests, is a multicast-enabled variant of LDA, the successor of which was described in Section 4.5.1. We have already seen that LDA+ is fair towards TCP while maintaining a smoother rate; it is equation based and utilizes 'packet pair' (see Section 4.6.3) to enhance its adaptation method. MLDA is actually a hybrid scheme in that it supports layered data encoding with group membership and has the sender adjust its transmission rate at the same time. The latter function compensates for bandwidth mismatch from coarse adaptation granularity – if there are few layers that represent large bandwidth steps, the throughput attained by a receiver without such a mechanism can be much too low or too high.

The *Packet Pair Receiver-driven Cumulative Layered Multicast (PLM)* scheme (Legout and Biersack 2000) is another notable approach; much like (Keshav 1991a), it is based upon 'packet pair' and the assumption of fair queuing in routers (Legout and Biersack 2002). *Fair Layered Increase/Decrease with Dynamic Layering (FLID-DL)* (Byers et al. 2000) is a generalization of RLC; by using a 'digital fountain' encoding scheme at the source, receivers are enabled to decode the original data once they have received a certain number of arbitrary but distinct packets. This renders the scheme much more flexible than other layered multicast congestion control proposals. Layers are dynamic in FLID-DL: their rates change over time. This causes receivers to automatically reduce their rates unless they join additional layers – thus, the common problem with long latencies that occur when receivers want to leave a group is solved. As with RLC, joining groups happens in a synchronized manner. While FLID-DL is, in general, a considerable improvement over RLC, it is not without faults: (Widmer 2003) points out that, just like RLC, it does not take the RTT into account, and this may cause unfair behaviour towards TCP under certain conditions.

Unlike MLDA, both RLC and FLID-DL do not provide feedback to the sender. Neither does *Wave and Equation-Based Rate Control (WEBRC)* (Luby et al. 2002), but this scheme has the notion of a 'multicast round-trip time' (MRTT) (as opposed to the unicast RTT), which is measured as the delay between sending a 'join' and receiving the first corresponding packet. WEBRC is a fairly complex, equation-based protocol that has the notion of 'waves' – these are used to convey reception channels that have a varying rate. In addition, there is a base channel that does not fluctuate as wildly as the others. A wave consists of a bandwidth aggregate from the sender that quickly increases to a high peak value and exponentially decays; this reduces the join and leave latency. WEBRC was specified as a 'building block' for reliable multicast transport in the RFC 3738 (Luby and Goyal 2004).

4.6   Better-than-TCP congestion control

Congestion control in TCP has managed to maintain the stability of the Internet while allowing it to grow the way it did. Despite this surprising success, these mechanisms are quite old now, and it would in fact be foolish to assume that finding an alternative method that simply works better is downright impossible (see Section 6.1.2 for some TCP criticism). Moreover, they were designed when the infrastructure was slightly different and, in general, a bit less heterogeneous – now, we face a diverse mixture of link layer technologies, link speeds and routing methods (e.g. asymmetric connections) as well as an immense variety of applications, and problems occur. This has led researchers to develop a large number of alternative congestion control mechanisms, some of which are incremental TCP improvements, while others are the result of starting from scratch; there are mechanisms that rely on additional implicit feedback, and there are others that explicitly require routers to participate.

One particular problem that most of the alternative proposals are trying to solve is the poor behaviour of TCP over LFPs. Figure 4.9, which depicts a TCP congestion-avoidance mode 'sawtooth' with link capacities of c and $2c$, shows what exactly this problem is: the area underneath the triangles represents the amount of data that is transferred. Calculating the area in (a) yields $3ct$ whereas the area in (b) gives $6ct$ – this is twice as much, just like the link capacity, and therefore the relative link utilization stays the same. Even then, the time it takes to fully saturate the link is also twice as long; this can become a problem in practice, where there is more traffic than just a single TCP flow and sporadic packet drops can prevent a sender from ever reaching full saturation.

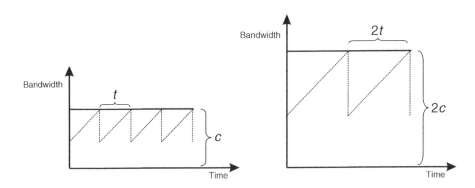

Figure 4.9   TCP congestion avoidance with different link capacities

The relationship between the packet loss ratio and the achievable average congestion window can also be deduced from Equation 3.6, which can be written as

$$T = \frac{1.2s}{RTT\sqrt{p}} \tag{4.8}$$

In order to fill a link with bandwidth T, the window would have to be equal to the product of RTT and T in the equation above, which requires the following packet loss probability p:

$$p = \left(\frac{1.2s}{T * RTT}\right)^2 \tag{4.9}$$

and therefore, the larger the link bandwidth, the smaller the packet loss probability has to be (Mascolo and Racanelli 2005). This problem is described as follows in RFC 3649 (Floyd 2003):

> The congestion control mechanisms of the current Standard TCP constrains the congestion windows that can be achieved by TCP in realistic environments. For example, for a Standard TCP connection with 1500-byte packets and a 100 ms round-trip time, achieving a steady-state throughput of 10 Gbps would require an average congestion window of 83,333 segments, and a packet drop rate of at most one congestion event every 5,000,000,000 packets (or equivalently, at most one congestion event every 1 2/3 hours). This is widely acknowledged as an unrealistic constraint.

The basic properties of the TCP AIMD behaviour are simply more pronounced over LFPs: the slow increase and the fact that it must (almost, with ECN) overshoot the rate in order to detect congestion and afterwards reacts to it by halving the rate. Typically, a congestion control enhancement that diverges from these properties will therefore work especially well over LFPs – this is just a result of amplifying its behaviour. In what follows, some such mechanisms will be described; most, but not all, of them were designed with LFPs in mind, yet their advantages generally become more obvious when they are used over such links. They are roughly ordered according to the amount and type of feedback they use, starting with the ones that have no requirements in addition to what TCP already has and ending with two mechanisms that use fine-grain explicit feedback.

4.6.1 Changing the response function

HighSpeed TCP and Scalable TCP

The only effort that was published as an RFC – *HighSpeed TCP*, specified in RFC 3649 (Floyd 2003) – is an experimental proposal to change the TCP rate update *only* when the congestion window is large. This protocol is therefore clearly a technology that was designed for LFPs only – the change does not take effect when the bottleneck capacity is small or the network is heavily congested. Slow start remains unaltered, and only the sender is modified. The underlying idea of HighSpeed TCP is to change *cwnd* in a way that makes it possible to achieve a high window size in environments with realistic packet loss ratios. As with normal TCP, an update is carried out whenever an ACK arrives at the sender; this is done as follows:

$$Increase : cwnd = cwnd + a(cwnd)/cwnd \qquad (4.10)$$

$$Decrease : cwnd = (1 - b(cwnd)) * cwnd \qquad (4.11)$$

where $a(cwnd)$ and $b(cwnd)$ are functions that are set depending on the value of $cwnd$. In general, large values of $cwnd$ will lead to large values of $a(cwnd)$ and small values of $b(cwnd)$ – the higher the window, the more aggressive the mechanism becomes.

TCP-like behaviour is given by $a(cwnd) = 1$ and $b(cwnd) = 0.5$, which is the result of these functions when $cwnd$ is smaller or equal to a constant called *Low_Window*. By default, this constant is set to 38 MSS – sized segments, which corresponds to a packet drop rate of 10^{-3} for TCP. There is also a constant called *High_Window*, which specifies the upper end of the response function; this is set to 83,000 segments by default (which is roughly the window needed for the 10 Gbps scenario described in the quote on Page 161) and another constant called *High_P*, which is the packet drop rate assumed for achieving a $cwnd$ of *High_Window* segments on average. *High_P* is set to 10^{-7} in RFC 3649 as a reasonable trade-off between loss requirements and fairness towards standard TCP. Finally, a constant called *High_Decrease* limits the minimum decrease factor for the *High_Window* window size – by default, this is set to 0.1, which means that the congestion window is reduced by 10%. From all these parameters, and with the goal of having $b(cwnd)$ vary linearly as the log of $cwnd$, functions that yield the results of $a(cwnd)$ and $b(cwnd)$ for congestion windows between *Low_Window* and *High_Window* are derived in RFC 3649. The resulting response function additionally has the interesting property of resembling the behaviour shown by a number of TCP flows at the same time, and this number increases with the window size.

The key to the efficient behaviour of HighSpeed TCP is the fact that it updates $cwnd$ with functions of $cwnd$ itself; this leads to an adaptation that is proportional to the current rate of the sender. This protocol is, however, not unique in this aspect; another well-known example is *Scalable TCP* (Kelly 2003), where the function $b(cwnd)$ would have the constant result 1/8 and the window is simply increased by 0.01 if no congestion occurred. Assuming a receiver that delays its ACKs, this is the same as setting $a(cwnd)$ to $0.005 * cwnd$ according to RFC 3649, which integrates this proposal with HighSpeed TCP by describing it as just another possible response function.

Scalable TCP has the interesting property of decoupling the loss event response time from the window size: while this period depends on the window size and RTT in standard TCP, it only depends on the RTT in the case of Scalable TCP. In Figure 4.9, this would mean that the sender requires not $2t$ but t seconds to saturate the link in (b), and this is achieved by increasing the rate exponentially rather than linearly – note that adding a constant to $cwnd$ is also what a standard TCP sender does in slow start. Scalable TCP just uses a smaller value.

HighSpeed TCP is only one of many proposals to achieve greater efficiency than TCP over LFPs. What makes it different from all others is the fact that it is being pursued in the IETF; this is especially interesting because it indicates that it might actually be acceptable to deploy such a mechanism provided that the same precautions are taken:

- Only diverge from standard TCP behaviour when the congestion window is large, that is, when there are LFPs and the packet loss ratio is small.

- Do not behave more aggressively than a number of TCP flows would.

Any such endeavour would have to be undertaken with caution; RFC 3649 explicitly states that decisions to change the TCP response function should not be made as an individual *ad hoc* decision, but in the IETF.

BIC and CUBIC

In (Xu et al. 2004), simulation studies are presented that show that the common unfairness of TCP with different RTTs is aggravated in HSTCP and STCP.[14] This is particularly bad in the presence of normal FIFO queues, where phase effects can cause losses to be highly synchronized – STCP flows with a short RTT can even completely starve off ones that have a longer RTT. On the basis of these findings, the design of a new protocol called *Binary Increase TCP (BI-TCP)* is described; this is now commonly referred to as *BIC-TCP* or simply *BIC*. By falling back to TCP-friendliness as defined in Section 2.17.4 when the window is small, BIC is designed to be gradually deployable in the Internet just like HSTCP.

This mechanism increases its rate like a normal TCP sender until it exceeds a pre-defined limit. Then, it continues in fixed size steps (explained below) until packet loss occurs; after that, it realizes a binary search strategy based on a maximum and minimum window. The underlying idea is that, after a typical TCP congestion event and the rate reduction thereafter, the goal is to find a window size that is somewhere between the maximum (the window at which packet loss occurred) and the minimum (the new window). Binary search works as follows in BIC: A midpoint is chosen and assumed to be the new minimum if it does not yield packet loss. Otherwise, it is the new maximum. Then, the process is repeated, until the update steps are so small that they would fall underneath a pre-defined threshold and the scheme has converged. BIC converges quickly because the time it takes to find the ideal window with this algorithm is logarithmic.

There are of course some issues that must be taken into consideration: since BIC is designed for high-speed networks, its rate jumps can be quite drastic – this may cause instabilities and is therefore constrained with another threshold. That is, if the new midpoint is too far away from the current window, BIC additively increases its window in fixed size steps until the distance between the midpoint and the current window is smaller than one such step. Additionally, if the window grows beyond the current maximum in this manner, the maximum is unknown, and BIC therefore seeks out the new maximum more aggressively with a slow-start procedure; this is called 'max probing'.

In (Rhee and Xu 2005), a refinement of the protocol by the name *CUBIC* is described. The main feature of this updated variant is that its growth function does not depend on the RTT; this is desirable when trying to be selectively TCP-friendly in case of little loss only because it allows to precisely detect such environment conditions. The dependence of HSTCP and STCP on *cwnd* (which depends not only on the packet loss ratio but also on the RTT) enables these protocols to act more aggressively than TCP when loss is significant but the RTT is short. CUBIC is the result of searching for a window growth

---

[14]These are common abbreviations for HighSpeed TCP and Scalable TCP, and we will use them from now on.

function that retains the strengths of BIC yet simplifies its window control and enhances its TCP-friendliness. As the name suggests, the function that was found is cubic; its input parameters are the maximum window size (from normal BIC), a constant scaling factor, a constant multiplicative decrease factor and the time since the last drop event. The fact that this 'real' time dominates the control is what makes it independent of the RTT: for two flows that experience loss at the same time, the result of this function depends on the time since the loss occurred even if they have different RTTs.

The cubic response function enhances the TCP-friendliness of the protocol; at the same time, the fact that the window growth function does not depend on the RTT allows TCP to be more aggressive than CUBIC when the RTT is small. Therefore, CUBIC additionally calculates how fast a TCP sender would open its window under similar circumstances and uses the maximum of this result and the result given by the standard CUBIC function.

TCP Westwood+

TCP Westwood+ (a refined version of 'TCP Westwood', first described in (Grieco and Mascolo 2002)) resembles HSTCP and STCP in that it changes the response function to react not in a fixed manner but in proportion to the current state of the system. It is actually quite a simple change, but there is also a fundamental difference: in addition to the normal inputs of a TCP response function, TCP Westwood+ utilizes the actual rate at which packets arrive at the receiver as an input – this is determined by monitoring the rate of incoming ACKs. Strictly speaking, this protocol therefore uses slightly more implicit feedback than the ones discussed so far. A result of this is that it works well over wireless networks: it will reduce its rate severely only when a significant number of packets are lost, which may either indicate severe congestion or a long series of corruption-based loss events. A single loss event that is due to link noise does not cause much harm – this is different from standard TCP. Also, while this protocol is TCP-friendly according to (Grieco and Mascolo 2004), it does not distinguish between 'low loss' and 'high loss' scenarios.

The response function of TCP Westwood+ is easily explained: its only divergence from standard TCP Reno or NewReno behaviour is the update of *ssthresh* (the starting point for fast recovery) in response to a congestion event. Instead of applying the fixed rate decrease by half, it sets this variable to the product of the estimated bandwidth (as determined from the rate of ACKs) and the minimum RTT – this leads to a more-drastic reduction in a case of severe congestion than in a case of light congestion. The key to the efficiency of TCP Westwood+ is the function that counts and filters the stream of incoming ACKs; this was designed using control theory. While the original version of TCP Westwood was prone to problems from 'ACK compression' (see Section 4.7), this was solved in TCP Westwood+ by means of a slightly modified bandwidth estimation algorithm.

4.6.2 Delay as a congestion measure

When congestion occurs in the network, a queue grows at the bottleneck and eventually, if the queue length cannot compensate, packets will be dropped. It therefore seems reasonable to react upon increasing delay. Note that only changes can be used, as a fixed delay measure does not explain much: two mildly congested queues may yield the same delay as a single severely congested one. Only delay *changes* can be used, that is, measurements must always be (implicitly or explicitly) combined with a previously recorded value. Delay measurements already control the behaviour of TCP via its RTT estimate, and increasing delay (which might be caused by congestion) will even make it react slower; still, the measurements are only interpreted for the sake of proper self-clocking and not as a real congestion indication.

Using delay in this manner has the potential advantage that a congestion control mechanism can make a more reasonable decision because it has more feedback. At the same time, it is non-intrusive, that is, it does not require any additional packets to be sent, and routers do not have to carry out extra processing. However, this feedback must be used with care, as it is prone to misinterpretations (see Section 2.4) – delay can also be caused by routing or link layer ARQ.

The idea of delay-based congestion control is not a new one; in fact, a mechanism that does this was described as early as 1989 (Jain 1989). In this scheme, the previous and current window and delay values are used to calculate a so-called normalized delay gradient, upon which the decision whether to increase or decrease the rate is based. Since then, several refinements were proposed; we will look at two popular ones.

TCP Vegas

An implicit form of delay-based congestion control can be found in a well-known TCP variant called *TCP Vegas* (Brakmo et al. 1994). This is a sender-side TCP modification that conforms to the normal TCP Reno standard: the specification does not forbid sending *less* than a congestion window, and this is exactly what TCP Vegas does. Like TCP Westwood+, it determines the minimum RTT of all measurements. The expected throughput is then calculated as the current size of *cwnd* divided by the minimum RTT. Additionally, the actual throughput is calculated by sending a packet, recording how many bytes are transferred until reception of the corresponding ACK and dividing this number by the sample RTT; if the sender fully utilizes its window, this should be equal to *cwnd* divided by SRTT. The calculation is carried out every RTT, and the difference between the actual rate and the expected rate governs the subsequent behaviour of the control.

The expected rate minus the actual rate is assumed to be non-negative, and it is compared against two thresholds. If it is underneath the lower threshold, Vegas increases *cwnd* linearly during the next RTT because it assumes that there are not enough data in the network. If it is above the upper threshold (which means that the expected throughput is much lower than the actual sending rate), Vegas decreases *cwnd* linearly. The congestion window remains unchanged when the difference between the expected and the actual rate is between the two thresholds. In this way, it can converge to a stable point of operation – this is a major difference between TCP Vegas and TCP Reno, which always needs to exceed the available capacity in order to detect congestion. Vegas can detect incipient congestion and react early, and this is achieved by monitoring delay: if the sender fully utilizes its congestion window, the only difference between the actual and expected rate calculated by Vegas is given by the minimum RTT versus the most-recent RTT sample.

This mechanism is the core element of TCP Vegas; there are two more features in the protocol: (i) it uses the receipt of certain ACKs as a trigger to check whether the RTO timer expired (this is an enhancement over older implementations with coarse timers), and (ii) it only increases the rate every other RTT in slow start so as to detect congestion during the intermediate RTTs. Despite its many theoretical advantages, Vegas is hardly used in practice nowadays because it is less aggressive than TCP Reno; if it was used in the Internet, it would therefore be 'pushed aside' (Grieco and Mascolo 2004). The value of TCP Vegas lies mainly in its historic impact as the first major TCP change that showed better behaviour by using more (implicit) network feedback; notably, there has been some work on refining it in recent years (Choe and Low 2004; Hasegawa et al. 2000).

FAST TCP

FAST TCP is a mechanism that made the news; I remember reading about it in the local media ('DVD download in five seconds' – on a side note, this article also pointed out that FAST even manages to reach such speeds with *normal TCP packet sizes!*). The FAST web page[15] mentions that it helped to break some 'land speed records' (there is a regular such competition in conjunction with the annual 'Supercomputing' conference). While this kind of hype cannot convince a serious scientist, it is probably worth pointing out that this mechanism is indeed among the most-elaborate attempts to replace TCP with a more-efficient protocol in high-speed networks; it has undergone extensive analyses and real-life tests.

FAST has been called a *high-speed version of Vegas*, and essentially, this is what it is, as it also takes the relationship between the minimum RTT and the recently measured RTT into account. In (Jin et al. 2004), the authors argue that alternatives like HSTCP and STCP are limited by the oscillatory behaviour of TCP (among other things), which is an unavoidable outcome of binary feedback from packet loss. This is not changed by ECN because an ECN-marked packet is interpreted just like a single packet loss event. They make a point for delay-based approaches in high-speed environments: in such networks, queuing delay can be much more accurately sampled because loss events are very rare, and loss feedback, in general, has a coarse granularity. The use of delay can facilitate quick driving of the system to the desired point of operation, which is near the 'knee' and not the 'cliff' in Figure 2.2 and therefore leaves some headroom for buffering web 'mice'. Finally, the fine-grain delay feedback enables FAST to reach a stable state instead of a fluctuating equilibrium.

FAST as described in (Jin et al. 2004) is split into four components: 'Data Control' (which decides *which* packets to transmit), 'Window Control' (which decides *how many* packets to transmit), 'Burstiness Control' (which decides *when* to transmit packets) and 'Estimation', which drives the other parts. Window Control and Burstiness Control operate at different timescales – in what follows, we are concerned with Estimation and Window Control, which makes decisions on the RTT timescale.

Estimation: From every ACK that arrives, the Estimation component takes an RTT sample and updates the minimum RTT and average RTT, which is calculated with an EWMA process. Other than standard TCP, FAST uses a weight for this process that is not a constant but depends on the current window size – roughly, the larger the window, the smaller the weight, that is, the smaller the influence of the most-recent sample. Also, this weight is usually much smaller than with TCP Reno. The average queuing delay is then calculated by subtracting the minimum RTT that was observed so far from the average RTT. Additionally, this component informs the other components about loss as indicated by the reception of three DupACKs.

---

[15]http://netlab.caltech.edu/FAST/

Window control: On the basis of the data from the 'Estimation' component, the window is updated using a single rule regardless of the state of the sender – there is, for instance, no distinction between rules to use when a packet was dropped as opposed to rules that must be applied when a non-duplicate ACK arrived. In every other RTT, the window is essentially updated as $w = w \frac{baseRTT}{RTT} + \alpha(w, qdelay)$, where w is the window, *baseRTT* is the minimum RTT, *RTT* is the average RTT and α is a function of the current window size and the average queuing delay. There is also a rule to prevent the window from being more than doubled in one update. In the current prototype, α is a constant, which produces linear convergence when the queuing delay is zero according to (Jin et al. 2004).

FAST TCP was shown to be stable and to converge to weighted proportional fairness, assuming that all users have a logarithmic utility function, and it was shown to perform very well in numerous simulations and real-life experiments, thereby making a good case for the inclusion of additional fine-grain feedback in the TCP response function.

4.6.3 Packet pair

Packet Pair is an exception because it is only a measurement methodology and not a protocol; it is generally agreed upon in the research community (or at least the IETF) that this technique does not yield information that is reliable enough for integration in an Internet-wide standard. Still, it is highly interesting because the information that it can retrieve could be quite useful for congestion control purposes; there have been proposals to integrate this function in such a mechanism – see (Keshav 1991a), for example, and (Hoe 1996), where it is proposed as a means to initialize *ssthresh*. Nowadays, researchers seem to focus on refining the measurement methodology while keeping an eye on congestion control as a possible long-term goal; here is a brief look at its basic principles and historic evolvement, roughly based on (Barz et al. 2005).

Figure 4.10 shows what happens when packets are 'squeezed' through a bottleneck: the 'pipes' on the left and right side of the figure are links with a high capacity and the 'pipe' in the middle is a small capacity link. The shaded areas represent packets; since their number of bits is not reduced as they are sent through the bottleneck, the packets must spread out in time. Even if the packets arrived at a lower rate than shown in the figure, they would probably be enqueued at the bottleneck and sent across it back-to-back. In any case, they leave it at a rate that depends on the capacity of the bottleneck – that is, if packets have a size of x bits and the bottleneck has a rate of x bit/s, then there will be 1 s from the beginning of a packet until the beginning of the next one. Unless one of these packets later experiences congestion, this is how they reach the receiver. This fact was probably first described with quite a similar diagram in (Jacobson 1988), where it is also explained that this is the rate at which ACKs are sent back (this was before delayed ACKs), and thus the bottleneck drives the self-clocking rate of a TCP sender. By monitoring the delay between a pair of packets (a 'packet pair'), the service rate of the bottleneck link can theoretically be deduced.

The term 'packet pair'[16] was probably first introduced in (Keshav 1991a), where it was also formally analysed in support of a control theory – based congestion control approach

---

[16]This measurement method is often also referred to as the 'packet pair approach'; variants thereof are sometimes called *packet dispersion techniques*.

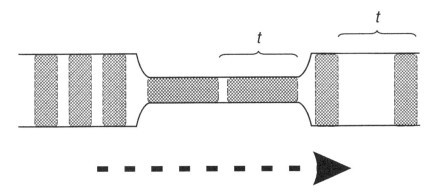

Figure 4.10 Packet pair

that assumed fair queuing in routers. An interesting Internet measurement study that does not use this name and actually relies on a large amount of UDP packets rather than a packet pair but still detects the bottleneck link capacity is documented in (Bolot 1993). A tool called *BProbe*, which uses packet pair in conjunction with ICMP ECHO packets, was described in (Carter and Crovella 1996); the authors used the tool to automatically colour links in a web site according to the speed of the corresponding bottleneck. In (Paxson 1999), an earlier version of which was published at ACM SIGCOMM in 1997, measurement studies from 20,000 TCP bulk data transfers between 35 Internet sites are presented, and a fundamental problem of the above approaches is outlined: since the measurements were carried out only at the sender, they are prone to errors from ACK compression – in other words, it is unclear whether the bottleneck was detected along the forward or backward path. Therefore, the measurements described in (Paxson 1999) were carried out at the receiver, where statistics was used to filter out errors. Sender-side path capacity estimation techniques were further criticized in (Allman and Paxson 1999), which, as already mentioned, made a point against using such measurements to find the initial value of *ssthresh*.

Van Jacobson wrote a tool called *pathchar*, that combines packet pair with traceroute-like functionality. An evaluation and refinement of the tool was described in (Downey 1999); like all packet pair – based measurement results, the output of this tool must be interpreted with care. The packet pair approach was refined in (Lai and Baker 2000) by sending a large packet immediately followed by a small one; this increases the chance of having the two packets serviced back-to-back by the bottleneck. In (Dovrolis et al. 2001), it is shown that the distribution attained with a large series of packet pairs is multimodal, that is, there can be several equally large peaks in the histogram; typically, a range of results is underneath the actual path bottleneck capacity and a range of results is above. Both effects can be caused by cross traffic, which could, for instance, cause the packet spacing to be increased after the bottleneck because only the second packet is enqueued. As shown in (Barz et al. 2005), if only the first packet is enqueued, this can lead to having the two packets sent back-to-back across a link that has a higher capacity than the bottleneck, yielding an estimate that is too high. One of the key findings in (Dovrolis et al. 2001) is that statistical techniques that emphasize the most-common bandwidth value or range can fail to yield the correct bottleneck link capacity. This is taken into account by a tool called *pathrate* that was presented in this document.

An idea called *packet triplet* is described in (Zixuan et al. 2003); here, not two but three back-to-back packets are sent, and the dispersions are compared so as to detect and filter out errors. *CapProbe* is yet another enhanced tool – it is superior to pathchar in accuracy and just as accurate as pathrate while providing faster estimation (Kapoor et al. 2004). CapProbe uses delay measurements to filter out errors; the underlying idea is that at least one of the two probing packets would be enqueued if the dispersion at the destination was distorted, and this is a result of cross traffic. The tool therefore uses the sample with the minimum delay sum (the cumulative delay of both packets) to estimate the bottleneck link capacity. CapProbe was successfully integrated in TCP and TFRC as a means to attain measurements in a passive (non-intrusive) manner by the authors of (Chen et al. 2004), and the fact that such passive measurements are feasible is underlined in (Jiang and Dovrolis 2004) by showing that about 50% of all packets from a standard TCP sender are packet pairs that were triggered by delayed ACKs.

An attempt to combine some of these recent developments is made in (Barz et al. 2005): measurements are taken at the TCP receiver, which is not prone to noise along the backward path and exploits the fact that ACKs can trigger a series of back-to-back packets from the sender. For a probe, the receiver window is used to limit the congestion window to three segments – this is done in order to have the sender generate packet triplets. The data are then filtered using the CapProbe delay filtering technique; it is shown that these two error-minimization methods nicely complement each other and enhance the chance of actually detecting the correct bottleneck capacity.

It should be clear from this discussion why packet dispersion techniques are not regarded as being reliable enough to be embedded in a congestion control mechanism in the IETF – as the numerous efforts to enhance the precision indicate, these methods are quite error-prone. On the other hand, the refinements are quite significant, and some recent results look promising; it is therefore unlikely that the research community will simply abandon the idea of integrating such measurements with congestion control in the near future.

4.6.4 Explicit feedback

As we have seen in the previous section, numerous efforts were carried out to enhance congestion control without requiring explicit help from the routers along a path – and some of them appear to work really well. There are several reasons to avoid using explicit feedback methods: packet processing is costly, especially in the core, and requiring routers to do extra work is therefore highly critical. Also, mechanisms such as explicit rate feedback (see Section 2.12) raise scalability concerns: what if every flow on the Internet requires routers to carry out extra work? Then, there is the deployment problem: how can we assume that all routers along an Internet path will be updated to support a new scheme, and what if only some of them do? These days, the focus in the research community is clearly on implicit feedback – based methods, and given all the potential problems above, this is quite understandable.

Explicit feedback also has some indisputable advantages: for one, congestion occurs in routers, so these devices simply have the most knowledge about it. Also, implicit feedback is always less precise – not just as a result of noise (cross traffic) but also because some things cannot be detected in such a way. Consider a single flow that traverses a 100 Mbps link at a rate of 50 Mbps, for example. How should this flow detect that it fills exactly half of the link capacity, and not 90 or 10%? If the link should be filled, the flow can therefore only blindly and cautiously increase its rate until it notices a change in its feedback. The earliest possible means to detect anything is increasing delay from queuing, but by then, the flow is already slightly above the limit. Therefore, it seems obvious that one can theoretically always do better than an implicit scheme with explicit feedback; the problems with such mechanisms are usually related to the feedback itself. Typically, the question on the table is: is this really worth the effort?

Multi-level ECN

Let us begin with an approach that is quite simple and also modest in terms of feedback requirements: *multi-level ECN*, which is described in (Durresi et al. 2001) among several other documents, describes another way of using the CE and ECT bits in the IP header. This idea was published before RFC 3168 (Ramakrishnan et al. 2001), which probably rendered it impractical because all bit combinations are now allocated in support of the ECN nonce. The remaining value of this experimental proposal is perhaps just that it shows how large the advantage that can be gained with a little more explicit feedback can be. In multi-level ECN, the bit combinations are assigned as follows:

$ECT = 0$, $CE = 0$: The sender is not ECN-capable; all other bit combinations indicate the contrary.

$ECT = 1$, $CE = 0$: There is no congestion.

$ECT = 0$, $CE = 1$: There is incipient congestion.

$ECT = 1$, $CE = 1$: There is moderate congestion.

In case of severe congestion, a router is supposed to drop the packet; this is a major difference between this mechanism and standard ECN, where packet drops and ECN-based congestion marks are assumed to indicate an equally severe congestion event.

Thus, in total, multi-level ECN supports three congestion levels. Marking in routers is carried out with an altered version of RED where an additional threshold called mid_{th} is introduced. If the average queue length is between min_{th} and mid_{th}, there is incipient congestion and the packet must be marked accordingly; if the average queue length is between mid_{th} and max_{th}, this is interpreted as moderate congestion. As with standard RED, packets are dropped only when the average queue length exceeds max_{th}. In order to inform the sender of the different congestion levels, the receiver must also make use of previously unused bit combinations of the CWR end ECE flags in the TCP header. The sender then reacts to packet drops as usual; its reaction to incipient congestion is to decrease *cwnd* by a certain fixed percentage, and the reaction to moderate congestion is to multiplicatively decrease *cwnd* with a factor that is greater than 0.5. Simulations indicate that even this quite simplistic way of using explicit feedback can already greatly enhance link utilization and reduce the average queue length as well as the number of packet drops.

Quick-start

HighSpeed TCP and most of its 'relatives' only address the response function but not the start-up mechanism, where a single sender initially does not know anything about the state of the network. Slow start can take a long time across LFPs, and it is quite an aggressive way to increase the rate given that not much is known; it was designed to merely bring the 'ACK clock' in motion. As discussed in Section 4.1.2, allowing a sender to blindly double its rate in a high-speed network may not be a good idea. Thus, according to RFC 3649 (Floyd 2003), starting with a higher window is a somewhat risky and structurally difficult problem that requires explicit support from all the routers along the path. This is separately addressed by a proposal called *Quick-start (QS)* (Jain and Floyd 2005), the idea of which is to use an IP option called *Quick-start Request* to ask routers whether they approve a requested rate (that is initially encoded in the option). Routers can approve, disapprove or reduce the rate to a value that is more suitable to them.

The QSR is carried in a SYN or SYN/ACK packet, and the answer is returned by the receiver in the next ACK with a 'Quick-start Response' option in the TCP header. In addition to a one-byte field that carries the requested rate (using a special encoding such as a table-based mapping or powers of two – this is currently undecided), the Quick-start Request option includes a TTL field that is initially set to a random number and decremented (modulo 256) by all routers that support the option. The receiver sends the difference between the TTL field in the IP header and the QS TTL field (modulo 256) back to the sender, where the same information is kept for comparison. The result is only used if the TTL difference in the packet equals the previously stored value because a different value means one of two things: (i) not all routers along the path support QS, or (ii) a malicious receiver tried to guess the correct value but failed. The latter function is similar in spirit to the ECN nonce.

There are several issues with QS: for instance, what if a router decides to delay (because of 'slow path' processing) or even drop a packet just because it contains an IP option? The latter is not an uncommon occurrence (Medina et al. 2005), and it is addressed in (Jain and Floyd 2005) by stating that the mechanism can still be useful in Intranets. Another open question is the previously mentioned common one for mechanisms that require explicit feedback: is it worth the effort? At the time of writing, QS is still in quite an early stage. Given the problems with slow start and lack of solutions to tackle them, it does, however, show promise.

CADPC/PTP

Yet another mechanism with a lengthy acronym, *Congestion Avoidance with Distributed Proportional Control (CADPC)*, relies much more strongly on explicit feedback than the schemes that were discussed so far. It was designed to show how far one can go with *nothing but* occasional fine-grain explicit rate feedback (see Section 2.12.2) while ensuring that its operation is scalable. The scheme consists of two parts: the CADPC congestion control mechanism and the *Performance Transparency Protocol (PTP)*, which is the signalling protocol that is used to retrieve the necessary feedback. Data packets are sent independently, for example, with UDP. *CADPC/PTP* means that CADPC is used with PTP; the intention of writing it in this way is to indicate that the individual parts to the left and right of the slash can be replaced. That is, PTP can be used by other mechanisms and CADPC can rely on a different means to obtain the required information. When combined with CADPC, PTP works as follows:

- The sender generates a PTP packet, which is transmitted towards the receiver. This packet is layered on top of IP to facilitate detection in routers; additionally, the IP 'Router Alert' option from RFC 2113 (Katz 1997) should be used. The PTP packet contains a request for 'Available Bandwidth' information.

- A PTP-capable router that sees this packet is supposed to add the following information to it:

  - The address of the network interface
  - A timestamp
  - The nominal link bandwidth (the 'ifSpeed' object from the 'Management Information Base (MIB)' of the router, specified in RFC 1213 (McCloghrie and Rose 1991))
  - A byte counter (the 'ifOutOctets' or 'ifInOctets' object from the MIB of the router, specified in the RFC 1213)

  These data can be added by prolonging the packet, which has the disadvantage of requiring some extra effort; it is therefore also possible for a sender to reserve the necessary space in the packet beforehand. The packet size can be chosen large enough for all routers to add their data while not exceeding the IP fragmentation limit on standard Internet paths.

- At the receiver, two consecutive such packets are required to calculate the bandwidth that was available during the period between them. Then, the nominal bandwidth B, the traffic λ and the interval δ of the dataset that has the smallest available bandwidth are fed back to the sender.

This is only one particular way of using PTP; the protocol is designed to efficiently retrieve any kind of performance-related information (so-called performance parameters – these could also be the average bottleneck queue length, the MTU of the path or the maximum expected bit error ratio, for example) from the network. It is scalable and lightweight; the code to be executed in routers is reduced to the absolute minimum and does not involve any per-flow state – all calculations are done at end nodes. Since the entities of relevance for PTP are links and not routers, a single intermediate router that does not support the protocol can be compensated for by additionally taking the incoming link into account; missing routers are detected with a TTL-based mechanism that closely resembles the QS TTL field of the previous section.

CADPC has a long name because it is a distributed variant of the 'Congestion Avoidance with Proportional Control' (CAPC) ATM ABR switch mechanism (see Section 3.8.1). Roughly, while CAPC has the rate increase or decrease proportional to the amount by which the total network traffic is below or above a pre-defined target rate, CADPC does the same with the relationship between the available bandwidth in the network and the rate of a sender. This is based on PTP feedback only, which can be very rare; in simulations, one packet every four RTTs suffices to attain good results. Whenever a PTP packet arrives, the rate $x(t + 1)$ of a sender is calculated as

$$x(t + 1) = x(t)\left(2 - \frac{x(t)}{\mathcal{B}(t)} - \frac{\lambda(t)}{\mathcal{B}(t)}\right) \qquad (4.12)$$

The old rate, $x(t)$, can be arbitrarily small but it must not be zero. This equation is a form of logistic growth, which can be shown to have an asymptotically stable equilibrium point at $\mathcal{B} * n/(1 + n)$. This means that the rate of CADPC converges to only half the bottleneck capacity for only one flow, but it converges to $100/101$ times the bottleneck capacity for hundred flows. In other words, the point of convergence itself quickly converges to the bottleneck capacity as the number of flows grows. This was decided to be acceptable because it is very unlikely that only, say, two or three users would share a high-capacity link.

Figure 4.11(a) depicts the cumulative throughput of 10 CADPC and 10 TCP flows that were sending data across a 10 Mbps bottleneck link in a simulation. Only flows of one type were used at a time: CADPC is not TCP-friendly and can therefore not be mixed with such flows. The relative speed of convergence in comparison with TCP grows with the bottleneck link capacity – running the same simulation with a 100 Mbps scenario shows that the CADPC flows reach their rate long before the TCP flows do so. This makes CADPC yet another mechanism that is especially useful in high-speed networks. It also includes a start-up enhancement that lets a single flow act like a number of flows for a while; as the flow approaches its point of convergence, this number is reduced until it eventually becomes one. The start-up behaviour is shown in Figure 4.11(b).

CADPC has some interesting properties: it converges to a max – min fair rate allocation and, like CUBIC, does not depend on the RTT. This means that the converged rate can very easily be calculated from the number of flows in the network and the bottleneck capacity,

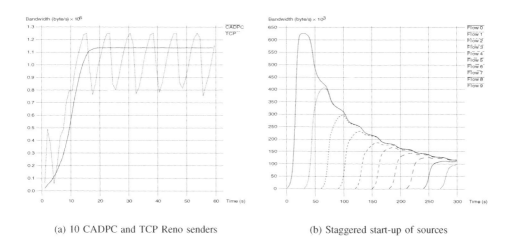

(a) 10 CADPC and TCP Reno senders   (b) Staggered start-up of sources

Figure 4.11 CADPC/PTP in action. Reproduced by kind permission of Springer Science and Business media

and it is also straightforward to realize flow differentiation by simply allowing one flow to act like a number of flows at the same time (two flows will obtain exactly twice the throughput of a single flow). Moreover, RTT independence means that losing or delaying a PTP packet will not cause much harm, as the convergence of the mechanism remains unaltered. CADPC achieves all this without any additional feedback – there are no ACKs from the receiver (which would of course be necessary to enable reliable data transmission with this protocol). The most comprehensive description of CADPC/PTP is (Welzl 2003); a shorter overview is given in (Welzl 2005c).

The eXplicit Control Protocol (XCP)

The *eXplicit Control Protocol (XCP)*, described in (Katabi 2003), is at the far end of the spectrum: by utilizing what might be the most amount of precise explicit feedback that can be used without risking scalability, the protocol achieves very high efficiency in a very wide range of scenarios. The feedback communication method is similar to explicit rate feedback as described in Section 2.12.2. It does not require per-flow state in routers, and unlike CADPC/PTP, it does not send additional signalling packets into the network. A key feature of this protocol is its decoupling of utilization control from fairness control – good utilization is achieved by adapting the rate on the basis of the spare bandwidth in the network and feedback delay, and fairness is attained by reclaiming bandwidth from flows that send more than their fair share.

In XCP, senders add a 'congestion header' that has three fields to data packets:

1. The current congestion window of the sender.

2. The current RTT estimate of the sender.

3. A feedback field that is initialized by the sender based on its demands and updated by routers; this is the only field that routers change – the others are read-only.

The sender updates its window on the basis of this feedback; additionally, per-packet ACKs ensure proper reaction to loss (which can be assumed to be rare in an XCP-controlled network). Routers monitor the input traffic to their output queues and use the difference between the bandwidth of a link and its input traffic rate to update the feedback value in the congestion header of packets. Thereby, the congestion window and RTT are used to do this in a manner that ensures fairness. By placing the per-flow state (congestion window and RTT) in packets like this, the mechanism avoids the additional burden of requiring per-flow state in routers.

XCP-capable routers compute the average RTT of the flows on a link from the RTT information in the congestion headers. A control decision is then made every average RTT; this is achieved by maintaining a timer for each link. The reasoning behind doing this every average RTT is that the router should wait for the senders to update their windows before making a change. The router logic is divided into two parts: an 'efficiency controller (EC)' and a 'fairness controller (FC)'. The EC computes the aggregate feedback as the number of bytes by which the traffic on a link should increase or decrease; this is a proportional rate update. Since the feedback would otherwise be zero when the link is saturated and the queue is persistently filled, the queue length is also taken into account.

Table 4.2   Differences between CADPC/PTP and XCP

Property	CADPC/PTP	XCP
Control	Rate based	Window based
Feedback	Extra signalling	In payload packets
Routers examine ...	Occasional signalling packets	Every packet
Convergence to ...	Capacity*flows/(flows+1)	Bottleneck capacity
Web background traffic	Does not work well	Works well

The FC realizes an AIMD policy: when the aggregate feedback allows the traffic to grow, per-packet feedback is computed in a way that has all flows equally increase their rate. This is attained by making the change of the congestion window proportional to the RTT of a flow; in order to translate this into per-packet feedback, the number of packets that the router expects to see from the flow during an average RTT must be computed. Combined with a concept called *bandwidth shuffling*, where the bandwidth is continuously redistributed according to the AIMD policy so as to maintain fairness, one ends up with a simple function of the RTT, packet size and congestion window. When the aggregate feedback requires the traffic volume to decrease, each flow should reduce its rate in proportion to its current throughput – this is attained by calculating per-packet negative feedback as a function of the RTT and packet size. The final XCP code is only three lines of code that must be executed whenever a packet arrives plus some simple calculations that are carried out whenever one of the link control timers fires.

XCP resembles CADPC/PTP in several ways: both protocols rely on fine-grain explicit feedback from all routers along a path, and their performance in simulations is somewhat similar (e.g. both show practically no loss and a very low average queue length). Yet, there are some quite important differences between the two – these are shown in Table 4.2. Each of them has its special strengths that make it more applicable than the other one in certain scenarios: whereas XCP reacts swiftly in the presence of short 'mice' flows, CADPC/PTP cannot properly react to network dynamics that occur on a much shorter timescale than the rate at which it sends its feedback packets. On the other hand, the effort for routers is less with CADPC/PTP, which also does not need any feedback messages in addition to the occasional PTP signalling packets. Conjointly, these protocols show that precise explicit feedback can be extremely beneficial for a congestion control mechanism.

4.6.5   Concluding remarks about better-than-TCP protocols

Coming up with congestion control that outperforms TCP, especially in high-speed environments, is an interesting research topic. It has recently gained momentum through proposals for protocols that are more aggressive than TCP yet deployable because they fall back to standard TCP behaviour when the loss ratio is high; additionally, there is an increasing demand for high-speed data transfer in Grid computing environments. Thus, obviously, there are many more efforts for congestion control that works better than TCP; a choice had to be made, and only a few seemingly representative ones could be described in this book. Some of the other recent developments include *H-TCP* (Shorten and Leith 2004), *TCP-Peach+* (Akyildiz et al. 2001, 2002), *LTCP* (Bhandarkar et al. 2005) and a number of UDP-based protocols including *Reliable Blast UDP (RBUDP)* and *Tsunami*; building a

congestion control mechanism on top of UDP greatly eases deployment, and this is facili-
tated by *UDT*, a library for building such protocols (Gu and Grossman 2005). An overview
of these and some other mechanisms is given in (Welzl et al. 2005).

One goal of this section was to show that researchers have gone to great lengths to
develop mechanisms that outperform TCP, and that most of these efforts seem to focus
on implicit feedback – and indeed, the performance that can be attained without requiring explicit help from routers is impressive. On the other hand, we have mechanisms like
XCP and CADPC/PTP that may perform even better but require router support; deployment of such schemes is a critical issue. For instance, both mechanisms are not inherently
TCP-friendly. It is suggested XCP be made interoperable by separately queuing XCP and
TCP flows in routers in (Katabi et al. 2002) – this could be done with CADPC/PTP just
as well, and there are already QoS mechanisms out there that can be used to realize such
separate queuing. With the good performance that these mechanisms yield, they could be
perfect candidates for an integrated QoS architecture as described in Section 6.2.4.

4.7   Congestion control in special environments

As mentioned at the beginning of this chapter, some of the proposed experimental enhancements in it could be regarded as mechanisms for special links; often, however, they are
beneficial in a diversity of environments, which is just a natural outcome of their properties.
CADPC/PTP, for instance, benefits from not relying on implicit feedback – all the related
problems that are explained in Section 2.4 are removed, including the possible misinterpretation of packet loss as a sign of congestion when it is actually caused by corruption.
Yet, it is (appropriately) listed among a series of mechanisms that are generally regarded
as enhancements for high-speed links in this chapter.

The intention of this section is to put things into perspective by summarizing how
certain environments and the mechanisms in this chapter go together.[17] Whether something
works well in a particular scenario simply depends on the properties of both elements: the
mechanism and the environment. Special links are briefly described in Section 2.13 – here
is a very quick overview of what their properties mean for a congestion control scheme:

Wireless networks: Packets can be dropped because of link noise, they can be delivered with
        errors and they can be delayed because of local retransmissions. All mechanisms that
        do not strictly rely on packet loss as the only (or most important) congestion indicator
        or can cope with erroneously delivered packets and/or sudden delay bursts will be
        beneficial in such environments. Examples are CADPC/PTP, which neither depends
        on packet loss nor on the RTT, CUBIC, which is also RTT independent, and the
        Eifel detection and response algorithms, which are designed to cope with sudden
        delay spikes.

Mobile networks: When hosts move, there can be sudden connection interruptions because
        a user might simply walk out of reach; handover mechanisms also take a while.

---

[17]Note that there are many specific TCP/IP enhancements for special links that are not discussed here because
they were not included in the book – for example, header compression or *ACK suppression* (selectively dropping
ACKs) in order to reduce congestion along the backward path in asymmetric networks (Samaraweera and Fairhurst
1998). Covering them would not make much sense in this chapter, which is broader in scope; a better source for
such things is (Hassan and Jain 2004). Also, the focus is solely on Chapter 4.

With IPv4, Mobile IP as specified in RFC 3344 (Perkins 2002) can cause packets to be forwarded from a foreign host via a 'home agent' (in the original network of the host) to the actual destination – this adds delay to the RTT. Moreover, packets can be sent back directly, and thus, the delay in the backward direction is usually shorter and the connection is asymmetric. Mechanisms that rapidly react to changing network conditions and are capable of quickly increasing their rate after a sudden interruption can be expected to be beneficial in such a scenario – Quick-Start would be one example. Also, the fact that the connection is asymmetric can have some implications, which will be discussed below.

LFPs: As explained in the previous section, LFPs tend to amplify the typical behaviour of a mechanism, and most TCP enhancements show more benefits when they are used in such environments than when the link capacity and delay are small. A problem of TCP that is normally not even noticed as being a problem is its potentially irregular packet spacing – this is problematic across LFPs, and pacing can help. Even though LFPs are common links, high bandwidth and long delay are actually two separate issues: mechanisms that have a more-efficient response function than TCP (e.g. HSTCP or STCP) work well across high capacity links, and mechanisms like CUBIC, which do not depend on the RTT, and connection splitting PEPs, which try to alleviate the adverse effects from long RTTs, work well across links with a long delay. All of these schemes will be beneficial across LFPs because they tackle at least one of the two problems.

Asymmetric networks: A connection is asymmetric when packets from host A to B traverse a different path than packets from B to A. This is generally not a big problem unless one of these paths is congested and the other is not. In particular, it is a common and problematic case to have a high-capacity link from a TCP sender to a TCP receiver and a very narrow link backwards – then, ACKs can cause congestion or be dropped, and TCP does not carry out congestion control on ACKs. When ACKs are enqueued back-to-back because there is congestion along the ACK path, this is an effect called *ACK compression* – and this can harm spacing of normal payload packets unless special care is applied. Mechanisms such as DCCP, which carry out congestion control for ACKs, are helpful in such environments, and it could also be useful to deploy a scheme that simply does not send much feedback (e.g. CADPC/PTP or LDA+). ACK compression can be a problem for a mechanism that measures the rate of incoming ACKs – and this was the case for TCP Westwood, but an update to the bandwidth estimation function in TCP Westwood+ solved the problem (Grieco and Mascolo 2002). Also, pacing can help because it avoids the immediate sending out of new data packets whenever an ACK arrives, and spurious timeout recovery may be beneficial because the delays of ACKs on the narrow return link can vary abruptly.

Table 4.3 gives a rough overview of some environments and shows which of the mechanisms in this chapter could be expected to be a good match. It is not comprehensive, and some relevant combinations may be missing – I tried to include only particularly suitable or typical combinations. For instance, AQM schemes eliminate phase effects, which may be more severe in the face of irregular packet spacing from asymmetric links (which may

Table 4.3   Applicability of mechanisms in this section for special environments

Mechanism	Wireless nets	LFPs	Asymmetric nets
ABC		×	
Limited slow start		×	
Congestion window validation		×	
Spurious timeout recovery	×	×	×
ETEN	×		
Separate checksums	×		
TCB interdependence, CM		×	
Connection splitting	×	×	
Pacing		×	×
Auto-tuning, DRS		×	
LDA+, TEAR		×	
DCCP	×		×
HSTCP, STCP, BIC		×	
CUBIC	×	×	
TCP Westwood+	×	×	
TCP Vegas, FAST TCP		×	
Multi-level ECN		×	
Quick-start		×	
CADPC/PTP	×	×	×
XCP	×	×	

cause the impact of an AQM scheme to be more pronounced), but that probably does not justify regarding an AQM mechanism as an enhancement for such links. Note that some real environments are in fact combinations of the above: for instance, a satellite network involves noise (which is often dealt with using FEC at the link layer) and is an LFP; at the same time, in a VSAT scenario, where a central hub transfers all data across a satellite and the endpoints communicate with the hub using some other technology (e.g. a regular telephone modem), the connection is highly asymmetric. Then, all of the problems (and solutions) for asymmetric networks as well as LFPs apply. Similarly, in Table 4.3, wireless networks are only considered to have only the properties in the list and not the additional link outages and asymmetric routing that is common in mobile links.

A highly recommendable reference on TCP interactions with all kinds of link layer technologies is (Hassan and Jain 2004). RFCs are also a good source for further information, as they typically omit link layer details that are often complicated yet irrelevant from a TCP/IP point of view. Here is a list of some RFCs that may be helpful when probing further:

Wireless links: RFC 3155 (Dawkins et al. 2001b), RFC 3366 (Fairhurst and Wood 2002)

Satellites: RFC 2488 (Allman et al. 1999a), RFC 2760 (Allman et al. 2000)

Asymmetric links: RFC 3449 (Balakrishnan et al. 2002)

A diverse range of link layer technologies is discussed in RFC 3819 (Karn et al. 2004), and RFC 3135 (Border et al. 2001) presents some where PEPs make sense. RFC 3481 (Inamura et al. 2003) is about 2.5G and 3G networks and therefore discusses issues regarding wireless and mobile links as well as asymmetry. Finally, RFC 3150 (Dawkins et al. 2001a) is about slow links.

5

Internet traffic management – the ISP perspective

So far, our observations focused on the 'global picture' of Internet stability, with operating-system developers, application programmers and end users as the main players. This is clearly only half of the story; ISPs earn money with Internet technology, and they are also concerned with efficient usage of network resources. It is crucial to understand that revenue dictates all their actions – for instance, if a choice has to be made between using a complicated mechanism to better use the available capacities on the one hand or simply install more bandwidth on the other, the cheaper path is usually taken. This often means increasing capacities because manpower is expensive and the administration effort of adding bandwidth is often quite negligible. These days, overprovisioning is a financially attractive solution for many ISP problems.

In this chapter, we will assume that capacities of an ISP are scarce, and the decision is *not* to overprovision; which way to go of course depends on the exact effort required, and some solutions are more complex than others. Also, there may be cases where an ISP is already pushing the envelope of a certain technology and increasing capacities further would mean replacing expensive equipment or adding (and not upgrading) links – which in turn can cause problems that must somehow be dealt with. There are some means for ISPs to manage the traffic in their network – for example, it can be decided to change IP routing a bit and relocate a certain fraction of the packets that congest a link elsewhere. This is called *traffic engineering*. Another option is to differentiate between, say, a high-class customer who should always experience perfect conditions and a low-class customer who must be able to cope with a some degree of congestion. This is called *Quality of Service (QoS)*. For such a scheme to be financially beneficial, high-class customers would of course have to be charged more – and for this, they would typically expect guarantees to work all across the globe. This was a popular Internet dream for quite some years, but it never really worked out, which may not be solely because of technical problems; today, the mechanisms for differentiating between more important and less important traffic classes remain available as yet another tool for ISPs to somehow manage their traffic.

Network Congestion Control: Managing Internet Traffic   Michael Welzl
© 2005 John Wiley & Sons, Ltd

Both traffic engineering and QoS are very broad topics, and they are usually not placed in the 'congestion control' category. As such, their relevance in the context of this book is relatively marginal. My intention is to draw a complete picture, and while these mechanisms are not central elements of congestion control, they cannot be entirely ignored. This chapter provides a very brief overview of traffic management; it is based upon (Armitage 2000) and (Wang 2001), which are recommendable references for further details.

5.1   The nature of Internet traffic

Before we plunge right into the details of mechanisms for moving Internet traffic around, it might be good to know what exactly we are dealing with. The traffic that flows across the links of an ISP has several interesting properties. First of all, it depends on the context – the time of day (in the middle of the night, there is typically less traffic than at morning peak hours, when everybody checks email), the number and type of customers (e.g. business customers can be expected to use peer-to-peer file sharing tools less often than private customers do) and the geographical position in the Internet. Second, there can be sudden strange occurrences and unexpected traffic peaks from viruses and worms. Not all of the traffic that traverses a link stems from customers who *want* to communicate – the authors of (Moore et al. 2001) even deduce worldwide denial-of-service activity from a single monitored link![1]

Despite the unexpected success of some applications and increasing usage of streaming video and VoIP, TCP makes up the majority of traffic in the Internet (Fomenkov et al. 2004). This means that we are dealing with an aggregate of flows that are mostly congestion controlled – unresponsive flows can indeed cause great harm in such a scenario. Moreover, this traffic is prone to all the peculiarities of TCP that we have seen in the previous chapters, that is, link noise is very problematic (it can lead to misinterpretations of corruption as a sign of congestion) and so is reordering (it can cause the receiver to send DupACKs, which makes the sender reduce its congestion window). The latter issue is particularly important for traffic management because it basically means that packets from a single TCP flow should not be individually routed – rather, they should stay together.

Most of the TCP traffic stems from users who surf the web. Web flows are often very short-lived – it is common for them to end before TCP even reaches *ssthresh*, that is, they often remain in their slow-start phase. Thus, Internet traffic is a mixture of long-lived data transfers (often referred to as *elephants*) and such short flows (often called *mice*) (Guo and Matta 2001). Web traffic has the interesting property of showing self-similarity (Crovella and Bestavros 1997) – a property that was first shown for Ethernet traffic in the seminal paper (Leland et al. 1993). This has a number of implications – most importantly, it means that some mathematical tools (traditional queuing theory models for analysing telephone networks) may not work so well for the Internet because the common underlying notion that all traffic is Poisson distributed is invalid. What remains Poisson distributed are user arrivals, not the traffic they generate (Paxson and Floyd 1995).

Internet traffic shows long-range dependence, that is, it has a heavy-tailed autocorrelation function. This marks a clear difference between this distribution and a random process: the

---

[1]This was done by detecting what tools were used for traffic generation and applying knowledge regarding the choice of destination IP addresses in these tools.

autocorrelation function of a Poisson distribution converges to zero. The authors of (Paxson and Floyd 1995) clearly explain what exactly this means: if Internet traffic would follow a Poisson process and you would look at a traffic trace of, say, five minutes and compare it with a trace of an hour or a day, you would notice that the distribution flattens as the timescale grows. In other words, it would converge to a mean value because a Poisson process has an equal amount of upward and downward motion. However, if you do the same with real Internet traffic, you may notice the same pattern at different timescales. When it may seem that a 10-min trace shows a peak and there must be an equally large dip if we look at a longer interval, this may not be so in the case of real Internet traffic – what we saw may in fact be a small peak on top of a larger one; this can be described as 'peaks that sit on ripples that ride on waves'. This recurrence of patterns is what is commonly referred to as self-similarity – in the case of Internet traffic, what we have is a self-similar time series.

It is well known that self-similarity occurs in a diverse range of natural, sociological and technical systems; in particular, it is interesting to note that rainfall bears some similarities to network traffic – the same mathematical model, a (fractional) autoregressive integrated moving average (fARIMA) process, can be used to describe both the time series (Gruber 1994; Xue et al. 1999).[2] The fact that there is no theoretic limit to the timescale at which dependencies can occur (i.e. you cannot count on the aforementioned 'flattening towards a mean', no matter how long you wait) has the unhappy implication that it may in fact be impossible to build a dam that is always large enough.[3] Translated into the world of networks, this means that the self-similar nature of traffic does have some implications on the buffer overflow probability: it does not decrease exponentially with a growing buffer size as predicted by queuing theory but it does so very slowly instead (Tsybakov and Georganas 1998) – in other words, large buffers do not help as much as one may believe, and this is another reason to make them small (see Section 2.10.1 for additional considerations).

What causes this strange property of network traffic? In (Crovella and Bestavros 1997), it was attributed to user think times and file size distributions, but it has also been said that TCP is the reason – indeed, its traffic pattern is highly correlated. This behaviour was called *pseudo–self-similarity* in (Guo et al. 2001), which makes it clear that TCP correlations in fact only appear over limited timescales. On a side note, TCP has been shown to propagate the self-similarity at the bottleneck router to end systems (Veres et al. 2000); in (He et al. 2002), this fact was exploited to enhance the performance of the protocol by means of mathematical traffic modelling and prediction. Self-similarity in network traffic is a well-studied topic, and there is a wealth of literature available; (Park and Willinger 2000) may be a good starting point if you are interested in further details.

No matter where it comes from, the phenomenon is there, and it may make it hard for network administrators to predict network traffic. Taking this behaviour into consideration in addition to the aforementioned unexpected possible peaks from worms and viruses, it seems wise for an ISP to generally overprovision the network and quickly do something when congestion is more than just a rare and sporadic event. In what follows, we will briefly discuss what exactly could be done.

---

[2] The stock market is another example – searching for 'ARIMA' and 'stock market' with Google yields some interesting results.

[3] This also has interesting implications on the stock market – theoretically, the common thinking 'the value of a share was low for a while, now it *must* go up if I just wait long enough' may only be advisable if you have an infinite amount of money available.

5.2 Traffic engineering

This is how RFC 2702 (Awduche et al. 1999) defines Internet traffic engineering:

> Internet traffic engineering is defined as that aspect of Internet network engineering dealing with the issue of performance evaluation and performance optimization of operational IP networks. Traffic Engineering encompasses the application of technology and scientific principles to the measurement, characterization, modelling, and control of Internet traffic.

This makes it clear that the term encompasses quite a diverse range of things. In practice, however, the goal is mostly routing, and we will restrict our observations to this core function in this chapter – from RFC 3272 (Awduche et al. 2002):

> One of the most distinctive functions performed by Internet traffic engineering is the control and optimization of the routing function, to steer traffic through the network in the most effective way.

Essentially, the problem that traffic engineering is trying to solve is the layer mismatch issue that was already discussed in Section 2.14: the Internet does not route around congestion. Congestion control functions were placed in the transport layer, and this is independent of routing – but ideally, packets should be routed so as to avoid congestion in the network and thereby reduce delay and packet loss. In mathematical terms, the goal is to *minimize the maximum* of link utilization. As mentioned before, TCP packets from a single end-to-end flow should not even be individually routed across different paths because reordering can cause the protocol to unnecessarily reduce its congestion window. Actually, such fast and dynamic routing would be at odds with TCP design, which is based upon the fundamental notion of a single pipe and not on an alternating set of pipes.

Why did nobody place congestion control into the network layer then? Traditionally, flow control functions *were* in the network layer (the goal being to realize reliability inside the network), and hop-by-hop feedback was used as shown in Figure 2.13 – see (Gerla and Kleinrock 1980). Because reliability is not a requirement for each and every application, such a mechanism does not conform with the end-to-end argument, which is central to the design of the Internet (Saltzer et al. 1984); putting reliability *and* congestion control into a transport protocol just worked, and the old flow control mechanisms would certainly be regarded as unsuitable for the Internet today (e.g. they probably do not scale very well). Personally, I believe that congestion control was not placed into the network layer because nobody managed to come up with a solution that works.

The idea of routing around congestion is not a simple one: say, path A is congested, so all traffic is sent across path B. Then, path B is congested, and it goes back to A again, and the system oscillates. Clearly, it would be better to send half of the traffic across path B and half of the traffic across path A – but can this problem to be solved in a way that is robust in a realistic environment? One problem is the lack of global knowledge. Say, a router decides to appropriately split traffic between paths A and B according to the available capacities at these paths. At the same time, another router decides to relocate some of its traffic along path B – once again, the mechanism would have to react. Note that we assume 'automatic routing around congestion' here, that is, the second router decided to use path B because another path was overloaded, and this of course depends on the congestion response of end systems. All of a sudden, we are facing a complicated system with all kinds of interactions, and the routing decision is not so easy anymore. This is not to say that automatizing traffic engineering is entirely impossible; for example, there is a related ongoing research project by the name of 'TeXCP'.[4]

Nowadays, this problem is solved by putting entities that have the necessary global knowledge into play: network administrators. The IETF defined tools (protocols and mechanisms) that enable them to manually[5] influence routing in order to appropriately fill their links. This, by the way, marks a major difference between congestion control and traffic management: the timescale is different. The main time unit of TCP is an RTT, but an administrator may only check the network once a day, or every two hours.

5.2.1   A simple example

Consider Figure 5.1, where the two PCs on the left communicate with the PC on the right. In this scenario, which was taken from (Armitage 2000), standard IP routing with RIP or OSPF will always select the upper path (across router D) by default – it chooses the shortest path according to link costs, and these equal 1 unless otherwise configured. This means that no traffic whatsoever traverses the lower path, the capacity is wasted and router D may unnecessarily become congested. As a simple and obvious solution to this problem that would not cause reordering within the individual end-to-end TCP flows, all the traffic that comes from router B could be manually configured to be routed across router C; traffic from router A would still automatically choose the upper path. This is, of course, quite a simplistic example – whether this method solves the problem depends on the nature and volume of incoming traffic among other things. It could also be a matter of policy: routers B and C could be shared with another Internet provider that does not agree to forward any traffic from router A.

How can such a configuration be attained? One might be tempted to simply set the link costs for the connection between the router at the 'crossroads' and router D to 2, that is, assign equal costs to the upper and the lower path – but then, all the traffic would still be

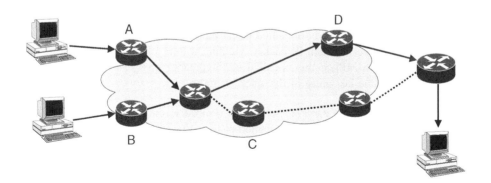

Figure 5.1 A traffic engineering problem

---

[4]http://nms.lcs.mit.edu/~dina/Texcp.html

[5]These things can of course also be automatized to some degree; for simplification, in this chapter, we only consider a scenario where an administrator 'sees' congestion and manually intervenes.

sent across only one of the two paths. Standard Internet routing protocols normally realize *destination-based routing*, that is, the destination of a packet is the only field that influences where it goes. This could be changed; if fields such as the source address were additionally taken into account, one could encode a rule like the one that is needed in our example. This approach is problematic, as it needs more memory in the forwarding tables and is also computation intensive.

IP in IP tunnelling is a simple solution that requires only marginal changes to the operation of the routers involved: in order to route all the traffic from router B across the lower path, this router simply places everything that it receives into another IP packet that has router B as the source address and router C as the destination address. It then sends the packets on; router C receives them, removes the outer header and forwards the inner packet in the normal manner. Since the shortest path to the destination is the lower path from the perspective of router C, the routing protocols do not need to be changed.

This mechanism was specified in RFC 1853 (Simpson 1995). It is quite old and has some disadvantages: it increases the length of packets, which may be particularly bad if they are already as large as the MTU of the path. In this case, the complete packet with its two IP headers must be fragmented. Moreover, its control over routing is relatively coarse, as standard IP routing is used from router B to C and whatever happens in between is not under the control of the administrator.

5.2.2   Multi-Protocol Label Switching (MPLS)

These days, the traffic engineering solution of choice is *Multi-Protocol Label Switching* (MPLS). This technology, which was developed in the IETF as a unifying replacement for its proprietary predecessors, adds a *label* in front of packets, which basically has the same function as the outer IP header in the case of IP in IP tunnelling. It consists of the following fields:

Label (20 bit): This is the actual label – it is used to identify an MPLS flow.

S (1 bit): Imagine that the topology in our example would be a little larger and there would be another such cloud in place of the router at the 'crossroads'. This means that packets that are already tunnelled might have to be tunnelled again, that is, they are wrapped in yet another IP packet, yielding a total of three headers. The same can be done with MPLS; this is called the emphlabel stack, and this flag indicates whether this is the last entry of the stack or not.

TTL (8 bit): This is a copy of the TTL field in the IP header; since the idea is not to require intermediate routers that forward labelled packets to examine the IP header, but TTL should still be decreased at each hop, it must be copied to the label. That is, whenever a label is added, TTL is copied to the outer label, and whenever a label is removed, it is copied to the inner label (or the IP header if the bottom of the stack is reached).

Exp (3 bit): These bits are reserved for experimental use.

MPLS was originally introduced as a means to efficiently forward IP packets across ATM networks; by enabling administrators to associate certain classes of packets with ATM

Virtual Circuits (VCs),[6] it effectively combines connection-oriented network technology with packet switching. This simple association of packets to VCs also means that the more-complex features of ATM that can be turned on for a VC can be reused in the context of an IP-based network. In addition, MPLS greatly facilitates forwarding (after all, there is only a 20-bit label instead of a more-complicated IP address), which can speed up things quite a bit – some core routers are required to route millions of packets per second, and even a pure hardware implementation of IP address based route lookup is slow compared to looking up MPLS labels. The signalling that is required to inform routers about their labels and the related packet associations is carried out with the *Label Distribution Protocol (LDP)*, which is specified in RFC 3036 (Andersson et al. 2001).

LDP establishes so-called *label-switched paths (LSPs)*, and the routers it communicates with are called *label-switching routers (LSRs)*. If the goal is just to speed up forwarding but not re-route traffic as in our example, it can be used to simply build a complete mesh of LSPs that are the shortest paths between all edge LSRs. Then, if the underlying technology is ATM, VCs can be set up between all routers (this is the so-called overlay approach' to traffic engineering) and the LSPs can be associated with the corresponding VCs so as to enable pure ATM forwarding. MPLS and LDP conjointly constitute a *control plane* that is entirely separate from the *forwarding plane* in routers; this means that forwarding is made as simple as possible, thereby facilitating the use of dedicated and highly efficient hardware. With an MPLS variant called *Multi-Protocol Lambda Switching (MPλS)*, packets can even be associated with a wavelength in all-optical networks.

When MPLS is used for traffic engineering, core routers are often configured to forward packets on the basis of their MPLS labels only. By configuring edge routers, multiple paths across the core are established; then, traffic is split over these LSPs on the basis of diverse selection criteria such as type of traffic, source/destination address and so on. In the example shown in Figure 5.1, the router at the 'crossroads' would only look at MPLS labels and router A would always choose an LSP that leads across router D, while router B would always choose an LSP that leads across router C. Nowadays, the speed advantage of MPLS switches over IP routers has diminished, and the ability to carry out traffic engineering and to establish tunnels is the primary reason for the use of MPLS.

5.3 Quality of Service (QoS)

As explained at the very beginning of this book, the traditional service model of the Internet is called *best effort*, which means that the network will do the best it can to send packets to the receiver as quickly as possible, but there are no guarantees. As computer networks grew, a desire for new multimedia services such as video conferencing and streaming audio arose. These applications were thought of as being workable only with support from within the network. In an attempt to build a new network that supports them via differentiated and accordingly priced service classes, ATM was designed; as explained in Section 3.8, this technology offers a range of services including ABR, which has some interesting congestion control-related properties.

---

[6]A VC is a 'leased line' of sorts that is emulated via time division multiplexing; see (Tanenbaum 2003) for further details.

The dream of bringing ATM services to the end user never really became a reality – but, as we know, TCP/IP was a success. Sadly, the QoS capabilities of ATM cannot be fully exploited underneath IP (although MPLS can now be used to 'revive' these features to some degree) because of a mismatch between the fundamental units of communication: cells and packets. Also, IP was designed not to make any assumptions about lower layers, and QoS specifications would ideally have to be communicated through the stack, from the application to the link layer in order to ensure that guarantees are never violated. A native IP solution for QoS had to be found.

5.3.1 QoS building blocks

The approach taken in the IETF is a modular one: services are constructed from somewhat independent logical building blocks. Depending on their specific instantiation and combination, numerous types of QoS architectures can be formed. An overview of the block types in routers is shown in Figure 5.2, which is a simplified version of a figure in (Armitage 2000). This is what they do:

Packet classification: If any kind of service is to be provided, packets must first be classified according to header properties. For instance, in order to reserve bandwidth for a particular end-to-end data flow, it is necessary to distinguish the IP addresses of the sender and receiver as well as ports and the protocol number (this is also called a *five-tuple*). Such packet detection is difficult because of mechanisms like packet fragmentation (while this is a highly unlikely event, port numbers could theoretically not be part of the first fragment), header compression and encryption.

Meter: A meter monitors traffic characteristics (e.g. 'does flow 12 behave the way it should?') and provides information to other blocks. Figure 5.3 shows one such mechanism: a *token bucket*. Here, tokens are generated at a fixed rate and put into a virtual 'bucket'. A passing packet 'grabs' a token; special treatment can be enforced

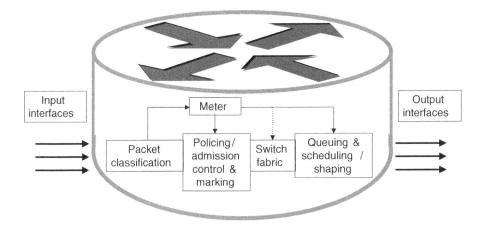

Figure 5.2 A generic QoS router

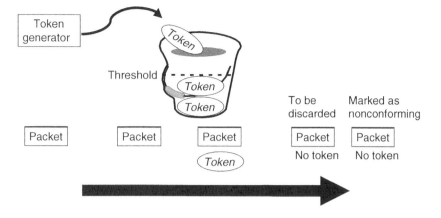

(a) Token bucket used for policing/marking

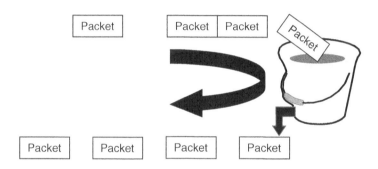

(b) Leaky bucket used for traffic shaping

Figure 5.3 Leaky bucket and token bucket

depending on how full the bucket is. Normally, this is implemented as a counter that is increased periodically and decreased whenever a packet arrives.

Policing: Under certain circumstances, packets are policed (dropped) – usually, the reason for doing so is to enforce conforming behaviour. For example, a limit on the burstiness of a flow can be imposed by dropping packets when a token bucket is empty.

Admission control: Other than the policing block, admission control deals with failed requirements by explicitly saying 'no'; for example, this block decides whether a resource reservation request can be granted.

Marking: Marking of packets facilitates their detection; this is usually done by changing something in the header. This means that, instead of carrying out the expensive *multi-field classification* process described above, packets can later be classified by simply looking at one header entry. This operation can be carried out by the router that marked the packet, but it could just as well be another router in the same domain. There can be several reasons for marking packets – the decision could depend on the conformance of the corresponding flow and a packet could be marked if it empties a token bucket.

Switch(ing) fabric: The switch fabric is the logical block where routing table lookups are performed and it is decided where a packet will be sent. The broken arrow in Figure 5.2 indicates the theoretical possibility of QoS routing.

Queuing: This block represents queuing methods of all kinds – standard FIFO queuing or active queue management alike. This is how discriminating AQM schemes which distinguish between different flow types and 'mark' a flow under certain conditions fit into the picture (see Section 4.4.10).

Scheduling: Scheduling decides when a packet should be removed from which queue. The simplest form of such a mechanism is a round robin strategy, but there are more complex variants; one example is *Fair Queuing (FQ)*, which emulates bitwise interleaving of packets from each queue. Also, there is its weighted variant *WFQ* and *Class-Based Queuing (CBQ)*, which makes it possible to hierarchically divide the bandwidth of a link.

Shaping: Traffic shapers are used to bring traffic into a specific form – for example, reduce its burstiness. A *leaky bucket*, shown in Figure 5.3, is a simple example of a traffic shaper: in the model, packets are placed into a bucket, dropped when the bucket overflows and sent on at a constant rate (as if there was a hole near the bottom of the bucket). Just like a token bucket, this QoS building block is normally implemented as a counter that is increased upon arrival of a packet (the 'bucket size' is an upper limit on the counter value), and decreased periodically – whenever this is done, a packet can be sent on. Leaky buckets enforce constant bit rate behaviour.

5.3.2   IntServ

As with ATM, the plan of the Integrated Services (IntServ) IETF Working Group was to provide strict service guarantees to the end user. The IntServ architecture includes rules to enforce special behaviour at each QoS-enabled network element (a host, router or underlying link); RFC 1633 (Braden et al. 1994) describes the following two services:

1. *Guaranteed Service (GS)*: this is for real-time applications that require strict bandwidth and latency guarantees.

2. *Controlled Load (CL)*: this is for elastic applications (see Section 2.17.2); the service should resemble best effort in the case of a lightly loaded network, no matter how much load there really is.

In IntServ, the focus is on the support of end-to-end applications; therefore, packets from each flow must be identified and individually handled at each router. Services are usually established through signalling with the Resource Reservation Protocol (RSVP), but it would also be possible to use a different protocol because the design of IntServ and RSVP (specified in RFC 2205 (Braden et al. 1997)) do not depend on each other. In fact, the IETF 'Next Steps In Signalling (NSIS)' working group is now developing a new signalling protocol suite for such QoS architectures.

5.3.3 RSVP

RSVP is a signalling protocol that is used to reserve network resources between a source and one or more destinations. Typically, applications (such as a VoIP gateway, for example) originate RSVP messages; intermediate routers process the messages and reserve resources, accept the flow or reject the flow. RSVP is a complex protocol; its details are beyond the scope of this book, and an in-depth description would perhaps even be useless as it might be replaced by the outcome of the NSIS effort in the near future. One key feature worth mentioning is *multicast* – in the RSVP model, a source emits messages towards several receivers at regular intervals. These messages describe the traffic and reflect network characteristics between the source and receivers (one of them, 'ADSPEC', is used by the sender to advertise the supported traffic configuration). Reservations are initiated by receivers, which send flow specifications to the source – the demanded service can then be granted, denied or altered by any involved network node. As several receivers send their flow specifications to the same source, the state is merged within the multicast tree.

While RSVP requires router support, it can also be tunnelled through 'clouds' of routers that do not understand the protocol. In this case, a so-called break bit is set to indicate that the path is unable to support the negotiated service. Adding so many features to this signalling protocol has the disadvantage that it becomes quite 'heavy' – RSVP is complex, efficiently implementing it is difficult, and it is said not to scale well (notably, the latter statement was relativized in (Karsten 2000)). RSVP traffic specifications do not resemble ATM style QoS parameters like 'average rate' or 'peak rate'. Instead, a traffic profile contains details like the token bucket rate and maximum bucket size (in other words, the burstiness), which refer to the specific properties of a token bucket that is used to detect whether a flow conforms.

5.3.4 DiffServ

Commercially, IntServ failed just as ATM did; once again, the most devastating problem might have been scalability. Enabling thousands of reservations via multi-field classification means that a table of active end-to-end flows and several table entries per flow must be kept. Memory is limited, and so is the number of flows that can be supported in such a way. In addition, maintaining the state in this table is another major difficulty: how should a router determine when a flow can be removed? One solution is to automatically delete the state after a while unless a refresh message arrives in time ('soft state'), but this causes additional traffic and generating as well as examining these messages requires processing power. There just seems to be no way around the fact that requiring information to be kept for each active flow is a very costly operation. To make things worse, IntServ routers do not only have to detect end-to-end flows – they also perform operations such as traffic shaping and scheduling on a per-flow basis.

The only way out of this dilemma appeared to be aggregation of the state: the *Differentiated Services (DiffServ)* architecture (specified in RFC 2475 (Blake et al. 1998)) assumes that packets are classified into separate groups by *edge routers* (routers at domain end-points) so as to reduce the state for inner (*core*) routers to a handful of classes; those classes are given by the *DiffServ Code Point (DSCP)*, which is part of the 'DiffServ' field in the IP header (see RFC 2474 (Nichols et al. 1998)). In doing so, DiffServ relies upon the aforementioned QoS building blocks. A DiffServ aggregate could, for instance, be composed of users that belong to a special class ('high-class customers') or applications of a certain type.

DiffServ comes with a terminology of its own, which was partially updated in RFC 3260 (Grossman 2002). An edge router that forwards incoming traffic is called *ingress routers*, whereas a router that sends traffic out of a domain is an *egress router*. The service between domains is negotiated using pre-defined *Service Level Agreements (SLAs)*, which typically contain non-technical things such as pricing considerations – the strictly technical counterpart is now called *Service Level Specification* (SLS) according to RFC 3260. The DSCP is used to select a *Per-Hop-Behaviour (PHB)*, and a collection of packets that uses the same PHB is referred to as a *Behaviour Aggregate* (BA). The combined functionality of classification, marking and possibly policing or rate shaping is called *traffic conditioning*; accordingly, SLAs comprise *Traffic Conditioning Agreements (TCAs)* and SLSs comprise *Traffic Conditioning Specifications (TCS)*.

Basically, DiffServ trades scalability for service granularity. In other words, the services defined by DiffServ (the most-prominent ones are *Expedited Forwarding* and the *Assured Forwarding* PHB Group) are not intended for usage on a per-flow basis; other than IntServ, DiffServ can be regarded as an incremental improvement on the 'best effort' service model. Since the IETF DiffServ Working Group started its work, many ideas based on DiffServ have been proposed, including refinements of the building blocks as above for use within the framework (e.g. the single rate and two rate 'three color markers' that were specified in RFC 2697 (Heinanen and Guerin 1999a) and RFC 2698 (Heinanen and Guerin 1999b), respectively).

5.3.5   IntServ over DiffServ

DiffServ is relatively static: while IntServ services are negotiated with RSVP on a per-flow basis, DiffServ has no such signalling protocol, and its services are pre-configured between edge routers. Users may want to join and leave a particular BA and change their traffic profile at any time, but the service is limited by unchangeable SLAs. On the other hand, DiffServ scales well – making it a bit more flexible while maintaining its scalability would seem to be ideal. As a result, several proposals for combining (i) the flexibility of service provisioning through RSVP or a similar, possibly more scalable signalling protocol with (ii) the fine service granularity of IntServ and (iii) the scalability of DiffServ have emerged; one example is (Westberg et al. 2002), and RFC 2998 (Paxson and Allman 2000) even specifies how to effectively run *IntServ over DiffServ*.

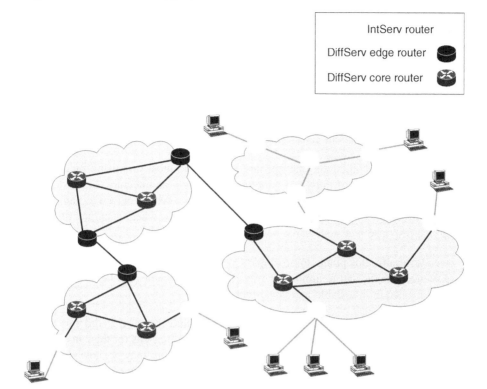

Figure 5.4 IntServ over DiffServ

No matter whether RSVP is associated with traffic aggregates instead of individual end-to-end flows (so-called *microflows*) or a new signalling protocol is used, the scenario always resembles the example depicted in Figure 5.4: signalling takes place between end nodes and IntServ edge routers or just between IntServ routers. Some IntServ-capable routers act like DiffServ edge routers in that they associate microflows with a traffic aggregate, with the difference that they use the IntServ traffic profile for their decision. From the perspective of an IntServ 'network' (e.g. a domain or a set of domains), these routers simply tunnel through a non-IntServ region. From the perspective of DiffServ routers, the IntServ network does not exist: packets merely carry the information that is required to associate them with a DiffServ traffic aggregate. The network 'cloud' in the upper left corner of the figure is such a DiffServ domain that is being tunnelled through, while the domain shown in the upper right corner represents an independent IntServ network. It is up to the network administrators to decide which parts of their network should act as DiffServ 'tunnels' and where the full IntServ capabilities should be used.

The IntServ/DiffServ combination gives network operators yet another opportunity to customize their network and fine-tune it on the basis of QoS demands. As an additional advantage, it allows for a clear separation of a control plane (operations like IntServ/RSVP signalling, traffic shaping and admission control) and a data plane (class-based DiffServ forwarding); this removes a major scalability hurdle.

5.4   Putting it all together

Neither IntServ nor DiffServ led to end-to-end QoS with the financial gain that was envisioned in the Industry – yet, support for both technologies is available in most commercial routers. So who turns on these features and configures them, and what for? These three quotes from RFC 2990 (Huston 2000) may help to put things into perspective:

> It is extremely improbable that any single form of service differentiation technology will be rolled out across the Internet and across all enterprise networks.

> The architectural direction that appears to offer the most promising outcome for QoS is not one of universal adoption of a single architecture, but instead use a tailored approach where scalability is a major design objective and use of per-flow service elements at the edge of the network where accuracy of the service response is a sustainable outcome.

> Architecturally, this points to no single QoS architecture, but rather to a set of QoS mechanisms and a number of ways these mechanisms can be configured to inter-operate in a stable and consistent fashion.

Some people regard Internet QoS as a story of failure because it did not yield the financial profit that they expected. There may be a variety of reasons for this; an explanation from RFC 2990 can be found on Page 212. Whatever the reasons may be, nowadays, RSVP, IntServ and DiffServ should probably be regarded as nothing but tools that can be useful when managing traffic. Whereas traffic engineering is a way to manage routing, QoS was conceived as a way to manage unfairness, but it is actually more than that: it is a means to classify packets into different traffic classes, isolate flows from each other and perform a variety of operations on them. Thus, the building blocks that were presented in Section 5.3.1 can be helpful even when differentiating between customers is not desired.

Constraint-based routing

RFC 2990 states that there is lack of a solution in the area of QoS routing; much like a routing algorithm that effectively routes around congestion, a comprehensive solution for routing based on QoS metrics has apparently not yet been developed. Both solutions have the same base problem, namely, the lack of global knowledge in an Internet routing protocol. Luckily, it turns out that traffic engineering and QoS are a good match. In MPLS, packets that require similar forwarding across the core are said to belong to a *forwarding equivalence class (FEC)* – this is a binding element between QoS and traffic engineering.

LDP establishes the mapping between LSPs and FECs. If the 'Exp' field of the label is not used to define FECs, these three bits can be used to encode a DiffServ PHB, that is, a combination of QoS building blocks such as queuing and scheduling that lead to a certain treatment of a flow.[7] This allows for a large number of options: consider a scenario where some important traffic is routed along link X. This traffic must not be subject to fluctuations,

---

[7]There are different possible encoding variants – the FEC itself could include the Exp field, and there could be several FECs that are mapped to the same LSP but require packets to be queued in a different manner.

that is, it must be protected from the adverse influence of bursts. Then, packets could first be marked (say, because a token bucket has emptied) and later assigned a certain FEC by a downstream router, which ensures that they are sent across path Y; in order to further avoid conflicts with the regular traffic along this path while allowing it to use at least a certain fraction of the bandwidth, it can be separately queued and scheduled with WFQ. All this can implicitly be encoded in the FEC.

This combination of the QoS building blocks in Section 5.3.1 and traffic engineering, where MPLS forwarding is more than just the combination of VCs and LSPs, is called *constraint-based routing* and requires some changes to LDP – for example, features like *route pinning*, that is, the specification of a fixed route that must be taken whenever a packet belongs to a certain FEC. As a matter of fact, not even traffic engineering as described with our initial example can be carried out with legacy MPLS technology because the very idea of sending all packets that come from router B across the lower path in Figure 5.1 embodies a constraint that can only be specified with an enhanced version of LDP that is called *CR-LDP*. A detailed explanation of CR-LDP and its usage can be found in RFC 3212 (Jamoussi et al. 2002) and RFC 3213 (Ash et al. 2002). RFC 3209 (Awduche et al. 2001) specifies a counterpart from the QoS side that can be used as a replacement for CR-LDP: RSVP with extensions for traffic engineering, *RSVP-TE*.

By enabling the integration of QoS building blocks with traffic engineering, CR-LDP (or RSVP-TE) and MPLS conjointly add another dimension to the flexibility of traffic management. Many things can be automatized, much can be done, and all of a sudden it seems that traffic could seamlessly be moved around by a network administrator. I would once again like to point out that this is not quite the truth, as the dynamics of Internet traffic are still largely governed by the TCP control loop; this significantly restrains the flexibility of traffic management.

Interactions with TCP

We have already discussed the essential conflict between traffic engineering and the requirement of TCP that packets should not be significantly reordered – but how does congestion control relate to QoS? Despite its age, RFC 2990 is still a rich source of information about the evolvement of and issues with such architectures; in particular, it makes two strong points about the combination of TCP and QoS:

1. If a TCP flow is provisioned with high bandwidth in one direction only, but there is congestion along the ACK path, a service guarantee may not hold. In particular, asymmetric routing may yield undesirable effects (see Section 4.7). One way to cope with this problem is to ensure symmetry, i.e. use the same path in both directions and amply provision it.

2. Traffic conditioning must be applied with care. Token buckets, for instance, resemble a FIFO queue, which is known to cause problems for congestion controlled flows (see Section 2.9). By introducing phase effects, it can diminish the advantages gained from active queue management; RFC 2990 even states that token buckets can be considered as 'TCP-hostile network elements'. Furthermore, it is explained that the operating stack of the end system would be the best place to impose a profile that is limited with a token bucket onto a flow.

After the explanation of the token bucket problem, the text continues as follows:

> The larger issue exposed in this consideration is that provision of some form of
> assured service to congestion-managed traffic flows requires traffic conditioning
> elements that operate using weighted RED-like control behaviours within the
> network, with less deterministic traffic patterns as an outcome.

This may be the most-important lesson to be learned regarding the combination of congestion control and QoS: *there can be no strict guarantees unless the QoS mechanisms within
the network take congestion control into account* (as is done by RED). One cannot simply assume that shaping traffic will lead to efficient usage of the artificial 'environment'
that is created via such mechanisms. TCP will always react in the same manner when
it notices loss, and it is therefore prone to misinterpreting any other reason for packet
drops – corruption, shaping and policing alike – as a sign of congestion.

Not everything that an ISP can do has a negative influence on TCP. AQM mechanisms,
for instance, are of course also under the control of a network provider. As a matter of fact,
they are even described as a traffic engineering tool that operates at a very short timescale
in RFC 3272 (Awduche et al. 2002) – and RED is generally expected to be beneficial even
if its parameters are not properly tuned (see Section 3.7). There are efforts to integrate QoS
with AQM that go beyond the simple idea of designing a discriminating AQM scheme
such as WRED or RIO – an example can be found in (Chait et al. 2002). The choice of
scheduling mechanisms also has an impact on congestion control; for example, on the
basis of (Keshav 1991a), it is explained in (Legout and Biersack 2002) that FQ can lead
to a much better 'paradigm' – this is just another name for what I call a 'framework' in
Section 6.1.2 – for congestion control.

In order to effectively differentiate between user and application classes, QoS mechanisms must protect one class from the adverse influences of the other. This fact can be
exploited for congestion control, for example, by separating unresponsive UDP traffic from
TCP or even separating TCP flows with different characteristics (Lee et al. 2001) (Laatu
et al. 2003). Also, the different classes can be provisioned with separate queues, AQM and
scheduling parameters; this may in fact be one of the most-beneficial ways to use QoS in
support of congestion control.

There are also ideas to utilize traffic classes without defining a prioritized 'high-class'
service; one of them is the *Alternative Best-Effort (ABE)* service, where there is a choice
between low delay (with potentially lower bandwidth) and high bandwidth (with potentially
higher delay), but no service is overall 'better' than the other. ABE provides no guarantees,
but it can be used to protect TCP from unresponsive traffic (Hurley et al. 2001). Other
proposals provide a traffic class *underneath* best-effort, that is, define that traffic is supposed
to 'give way' (Bless et al. 2003; Carlberg et al. 2001) – associating unresponsive traffic
with such a class may be an option worth investigating.

Consider the following example: an ISP provides a single queue at a bottleneck link
of 100 Mbps. Voice traffic (a constant bit rate UDP stream) does not exceed 50 Mbps,
and the rest of the available capacity is filled with TCP traffic. Overall, customers are
quite unhappy with their service as voice packets suffer large delay variations and frequent
packet drops. If the ISP configures a small queue size, the delay variation goes down, but
packet discards – both for voice and TCP – increase substantially. By simply separating
voice from TCP with DiffServ, providing the two different classes with queues of their

own and configuring a scheduler to give up to 50 Mbps to the voice queue and the rest to TCP, the problem can be solved; a small queue can be chosen for the voice traffic and a larger one with AQM for TCP. In order to make this relatively static traffic allocation more flexible, the ISP could allow the voice traffic aggregate to fluctuate and use RSVP negotiation with IntServ over DiffServ to perform admission control.

Admission control and congestion control

Deciding whether to accept a flow or reject it because the network is overloaded is actually a form of congestion control – this was briefly mentioned in Section 2.3.1, where the historically related problem of connection admission control in the telephone network was pointed out. According to (Wang 2001), there are two basic approaches to admission control: *parameter based* and *measurement based*. Parameter-based admission control is relatively simple: sources provide parameters that are used by the admission control entity in the network (e.g. an edge router) to calculate whether a flow can be admitted or not. These parameters can represent a fixed reserved bandwidth or a peak rate as in the case of ATM. This approach has the problem that it can lead to inefficient resource utilization when senders do not fully exploit the bandwidth that they originally specified. As a solution, the actual bandwidth usage can be measured, and the acceptance or rejection decision can be based upon this additional knowledge.

Dynamically measuring the network in order to tune the amount of data that are sent into the network is exactly what congestion control algorithms do – but admission control is carried out inside the network, and it is not done at the granularity of packets but connections, sessions (e.g. an HTTP 1.0 session that comprises a number of TCP connections which are opened and closed in series) or even users. In the latter case, the acceptance decision is typically guided by security considerations rather than the utilization of network resources, making it less relevant in the context of this book. Just like packet-based congestion control, measurement-based admission control can be arbitrarily complex; however, the most common algorithms according to (Wang 2001) – 'simple sum', 'measured sum', 'acceptance region' and 'equivalent bandwidth' – are actually quite simple. Measured sum, for instance, checks the requested rate of a flow plus the measured traffic load against the link capacity times a user-defined target utilization. The utilization factor is introduced in order to leave some headroom – at very high utilization, delay fluctuations will become exceedingly large, and this can have a negative impact on the bandwidth measurement that is used by the algorithm. An in-depth description and performance evaluation of the four basic algorithms can be found in (Jamin et al. 1997).

There is also the question of how to measure the currently used bandwidth. One approach is to use a so-called 'time window', where the average arrival rate is measured over a predefined sampling interval and the highest average from a fixed history of such sampling intervals is taken. Another method is to use an EWMA process, where the stability (or reactiveness) of the result is under the control of a fixed weighting parameter. How should one tune this parameter, and what is the ideal length of a measurement history in case of memory-based bandwidth estimation? Finding answers to these questions is not easy as they depend on several factors, including the rate at which flows come and go and the nature of their traffic; some such considerations can be found in (Grossglauser and Tse 1999).

Their inherent similarities allow congestion control and admission control to be integrated in a variety of ways. For instance, it has been proposed to use ECN as a decision element for measurement-based admission control (Kelly 2001). On the other side of the spectrum, there are proposals to enhance the performance of the network without assuming any QoS architecture to be in place. This is generally based on the assumption that the network as a whole becomes useless when there are too many users for a certain capacity. For instance, if the delay between clicking a link and seeing the requested web page exceeds a couple of seconds, users often become frustrated and give up; this leads to financial loss – it would be better to have a smaller number of users and ensure that the ones who are accepted into the network can be properly served.

In (Chen et al. 2001), a method to do this at a web server is described; this mechanism estimates the workload for a flow from the distribution of web object requests as a basis for the acceptance decision. By measuring the traffic on a link and reacting to TCP connection requests, for example, by sending TCP RST packets or simply dropping the TCP SYN, the loss rate experienced by the accepted TCP flows can be kept within a reasonable range (Mortier et al. 2000). The authors of (Kumar et al. 2000) did this in a non-intrusive manner by querying routers with SNMP, sniffing for packets on an Ethernet segment and occasionally rejecting flows with a TCP RST packet that has a spoofed source address and is sent in response to a TCP SYN. In their scenario, the goal was to control traffic according to different policies in a campus network – their system can, for instance, be used to attain a certain traffic mix (e.g. it can be ensured that SMTP traffic is not pushed aside by web surfers).

To summarize, depending on how such functions are applied, traffic management can have a negative or positive impact on a congestion control mechanism. Network administrators should be aware of this fact and avoid configuring their networks in ways that adversely affect TCP, as this can diminish the performance gains that they might hope to attain with tools such as MPLS. In Section 6.2.4, we will see that things can also be turned around: if effectively combined, a congestion control mechanism can become a central element of a QoS architecture where per-flow guarantees should be provided in a scalable manner.

6

The future of Internet congestion control

I wrote this chapter for Ph.D. students who are looking for new research ideas. It is a major departure from the rest of the book: here, instead of explaining existing or envisioned technology, I decided to provide you with a collection of my own thoughts about the future of Internet congestion control. Some of these thoughts may be quite controversial, and you might not subscribe to my views at all; this does not matter, as the goal is not to *inform* but to *provoke* you. This chapter should be read with a critical eye; if disagreement with one of my thoughts causes you to come up with a more reasonable alternative, it would have fulfilled its purpose. If you are looking for technical information, reading this chapter is probably a waste of time.

The Internet has always done a remarkable job at surprising people; technology comes and goes, and some envisioned 'architectures of the future' fail while others thrive. Neither the commercial world nor the research community foresaw the sudden success of the World Wide Web, and Internet Quality of Service turned out to be a big disappointment. This makes it very hard to come up with good statements about the future; congestion control can move in one direction or another, or it might not evolve at all over the next couple of years. We can, however, try to learn from 'success stories' and failures alike, and use this knowledge to step back and think again when making predictions. As always, *invariants* may help – these may be simple facts of life that did not seem to change over the years. To me, it seems that one of them is: 'most people do not want video telephony'. Just look at how often it has been tried to sell it to us – the first occurrence that I know of was ISDN, and just a couple of days ago I read in the news that Austrian mobile service providers are now changing their strategy because UMTS users do not seem to want video telephony but prefer downloading games, ring tones and music videos instead.

Another invariant is culture; customers in Asia and Europe sometimes want different services, and this fact is unlikely to change significantly within the next decade (in the world of computer networks, and computer science in general, this is a long time). To me,

Network Congestion Control: Managing Internet Traffic   Michael Welzl
© 2005 John Wiley & Sons, Ltd

it is a general invariant that people, their incentives and their social interactions and roles dictate what is used. The Internet has already turned from an academic network into a cultural melting pot; the lack of laws that are globally agreed upon and lack of means to enforce them rendered its contents and ways of using it extremely diverse. One concrete example is peer-to-peer computing: this area has significantly grown in importance in the scientific community. This was caused by the sudden success of file-sharing tools, which were illegal – who knows what all the Ph.D. students in this area would now be working on if the authorities had had a means to stop peer-to-peer file sharing. Heck, I am sitting in the lecture hall of the German 'Communication in Distributed Systems' (KiVS) conference as I write this,[1] and there will be an associated workshop just about peer-to-peer computing in this room tomorrow. Who knows if this workshop would take place if it was not for the preceding illegal activities?

The one key piece of advice that I am trying to convey here is: whatever you work on, remember that the main elements that decide what is used and how it is used are *people* and their *incentives*. People do not use what is designed just because researchers want them to; their actions are governed by incentives. Some further elaborations on this topic can be found in Section 2.16.

The contents of this chapter are not just predictions; they consist of statements about the current situation and certain ways of how things could evolve as well as my personal opinion about how they *should* evolve. Some parts may in fact read like a plea to tackle a problem. I hope that this mixture makes it a somewhat interesting and enjoyable read.

6.1 Small deltas or big ideas?

The scientific community has certainly had its share of 'small deltas' (minor improvements of the state of the art), especially in the form of proposed TCP changes. This is not to say that such research is useless: minor updates may be the most robust and overall sensible way to improve the things we use today. The sad part is that the vast majority of congestion control research endeavours falls in this category, and one has to search hard in order to find drastically new and entirely different ideas (one such example is the refreshing keynote speech that Van Jacobson gave at SIGCOMM 2001 – see Section 2.14.1). Let us explore the reasons that led to this situation.

Any research effort builds upon existing assumptions. In the case of Internet congestion control, some of them have led to rules for new mechanisms that are immensely important if any kind of public acceptance is desired – some examples are given below:

1. A mechanism must be scalable; it should work with an arbitrary number of users.

2. It must be stable, which should at least be proven analytically in the synchronous RTT, fluid traffic model case – other cases are often simulated.

3. It should be immediately deployable in the Internet, which means that it must be fair towards TCP (TCP-friendly).

These are certainly not all common rules (a good source for other reasonable considerations when designing new standards and protocols is RFC 3426 (Floyd 2002)), and indeed,

---

[1]There is a coffee break right now – I am *not* writing this during a talk!

they appear to make sense. Ensuring scalability cannot be a bad idea for a mechanism that is supposed to work in the Internet, and stability is obviously important. Rule three may not be as obvious as the first ones, but satisfying this criterion can certainly help to increase the acceptance of a mechanism, thereby making deployment more realistic.

On the other hand, research is about new ideas. Each and every constraint narrows the design space and can therefore prevent innovation. Sometimes, it may be a good idea to step back and rethink what has seemingly become common knowledge – new ideas should not always be abandoned just because a researcher wants to ensure that her mechanism satisfies all existing rules without questioning them. Doubtlessly, if everybody did that, none of the major inventions of the last century would have been possible. If we relax the constraints a bit, which may be appropriate in academic research, there is, in fact, a spectrum of viable approaches that may not always coincide with common design rules. Let us now see where rethinking one of the rules could lead us.

6.1.1 TCP-friendliness considerations

As we have seen in Chapter 4, the adoption of TCP congestion control did not mark the end of research in this area: since then, among many other things, active queue management (RED) and explicit communication between end nodes and the network (ECN) were introduced and more implicit information has been discovered (e.g. the bottleneck bandwidth, using packet pair) and used (e.g. in TCP Vegas or LDA+). However, in order for a mechanism to be gradually deployable in the Internet, it should still be compatible with TCP, which is the prevalent transport protocol. The idea is to protect existing TCP flows from ones that are unresponsive or too aggressive, as a single such flow can do severe harm to a large number of TCP flows. The original, and probably still most common, definition for a flow to be TCP-friendly ('TCP-compatible') can be found in Section 2.17.4; here it is again for your convenience:

> *A TCP-compatible flow is responsive to congestion notification, and in steady state it uses no more bandwidth than a conforming TCP running under comparable conditions.*

TCP causes congestion in order to avoid it: its rate is known to depend on the packet size, the RTT, the RTO (which is a function of instantaneous RTT measurements) and loss (Padhye and Floyd 2001). It reacts to changes of the RTT (usually due to changed queuing delay) and in response to loss.[2] Both of these factors obviously have a negative effect on the rate – and they are, to some degree, under control of the congestion control mechanism itself, which conflicts with the fact that they also represent the 'comparable conditions' in the definition.

Consider, as a simple thought experiment, a single flow of a certain type traversing a single link. This flow uses congestion control similar to TCP, but let us assume that it somehow knew the link capacity and would stop the additive-increase process in time (i.e. before its rate would reach the limit and incur increased queuing delay or loss). On average, this flow would send at a higher rate than would a TCP flow under comparable conditions and therefore not satisfy the constraints for TCP-friendliness.

---

[2] At this point, we neglect ECN for simplification.

Clearly, if there were two such flows, neither of them would be able to reach its limit and both would end up working just like TCP. Our mechanism would also act like TCP in the presence of a TCP flow – this means that its effect on TCP is similar to the effect of a TCP flow. Under certain conditions, our flow loses less and causes less queuing delay (less congestion); yet, as soon as loss and delay are under control of a TCP flow, it reduces the rate just like TCP would. As a matter of fact, our flow could even be TCP itself, if its maximum send window is tuned to the bandwidth × RTT product of the link. This means that even TCP itself is not always TCP-friendly, depending on its parameters! This parameter dependency is implicit in the definition, and it seems to be undesirable.

We can therefore argue that it should be a goal of new congestion control mechanisms to actually *avoid* congestion by reducing loss and queuing delay; if the response to increased queuing delay and loss is designed appropriately, it should still be possible to compete fairly with TCP *in the presence of TCP*. On the other hand, if they are not competing with TCP, it should be possible for such flows to work better than TCP – thereby creating an incentive to use them. Examples of mechanisms that might possibly be changed appropriately are XCP (Section 4.6.4) because it decouples fairness from congestion control and CADPC/PTP (Section 4.6.4) because it does not use packet loss feedback at all. To summarize, satisfying the common definition of TCP-friendliness is sufficient but not necessary for a flow to be TCP-friendly. The definition should instead be as follows:

> *A TCP-compatible flow is responsive to congestion notification, and its effect on competing TCP flows is similar to the effect of a TCP flow under comparable conditions.*

Since the idea of TCP-friendliness is merely to limit the negative impact on TCP flows, this new definition can be expected to be both necessary and sufficient.

It seems that the publication of HighSpeed TCP in RFC 3649 (Floyd 2003) already 'cleared the ground' for researchers to try and make their protocols *somewhat* TCP-friendly; as explained in Section 4.6.1, this protocol is only unfair towards TCP when the congestion window is very large, which also means that packet drop rates are very low. This is a situation where TCP is particularly inefficient, and the fact that HighSpeed TCP falls back to standard TCP behaviour when loss increases ensures that it cannot cause congestion collapse. Note that this approach bears some resemblance with the definition above – when the network is congested and many packets are lost, the protocol will act like TCP and therefore satisfy the definition, and when the packet drop rate is very low, the impact of a new protocol on TCP is generally not very large because TCP primarily reacts to loss.

Yet, while being more aggressive than TCP only when there is little loss may be a way to satisfy the new definition above, taking this method as the basis (i.e. adding '*when the loss ratio is small*' to the original definition) may be critical. For instance, the authors of (Rhee and Xu 2005) state that the regime where TCP performs well should not be defined by the window size but by the time between two consecutive loss events (called *congestion epoch* time). If the goal is to come up with a definition that lasts, this example shows that *adding details* is not a good approach. Instead, it might be better to focus on the essence of the TCP-friendliness idea – that TCP flows should not be harmed by others that abide by the definition.

6.1.2 A more aggressive framework

TCP is the predominant Internet transport protocol. Still, the embedded congestion control mechanisms show a significant number of disadvantages:

Stability: The primary goal of a congestion control mechanism is to bring the network to a somewhat stable state; in the case of TCP, control-theoretic reasoning led to some of the design decisions, but (i) it only reaches a fluctuating equilibrium, and (ii) the stability of TCP depends on several factors including the delay, network load and network capacity. In particular, as we have seen in Section 2.6.2, analysing the stability of the underlying AIMD control strategy is not as straightforward as it may seem in the case of asynchronous feedback delay (Johari and Tan 2001). Even then, the model may be quite an unrealistic simplification, as TCP encompasses many additional features that influence stability, most notably self-clocking. Figure 6.1 was obtained by plotting an 'ns' simulation result in a way that is similar to the vector diagrams in Section 2.5.1; the scenario consisted of two TCP Reno flows that were started at the same time and shared a single bottleneck link with FIFO queuing. The diagram shows the throughput obtained by the receivers. While perfectly synchronous operation is not possible when two packets are transferred across a single link or placed into a single queue, the diagram does not even remotely resemble the AIMD behaviour in Figure 2.4. When modelling TCP and not just AIMD, it can be shown that, locally, the system equilibrium can become unstable as delay or capacity

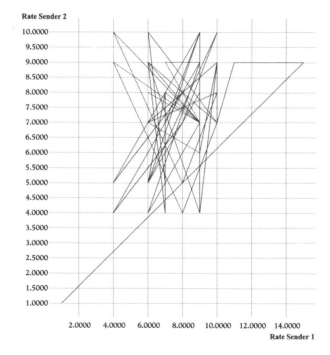

Figure 6.1 A vector diagram of TCP Reno. Reproduced by kind permission of Springer Science and Business Media

increases (Low et al. 2002). The results in (Vinnicombe 2002) additionally indicate that TCP is prone to instability when the congestion window and queuing delay are small.

Fairness: As discussed in Section 2.17.3, an ideal congestion control mechanism would probably maximize the respective utility functions of all users, as these functions represent their willingness to pay. One fairness measure that comes close to this behaviour (assuming 'normal' applications such as file downloads and web surfing) is proportional fairness – this was said to be realized by TCP, but, in fact, it is not (Vojnovic et al. 2000).

Heterogeneous links: When regular TCP is used across noisy (typically wireless) links, checksum failures due to transmission errors can cause packet drops that are misinterpreted as a congestion signal, thereby causing TCP to reduce its rate. With its fixed rate increase policy, the link utilization of TCP typically degrades as the link capacity or delay increases – it can take a long time to reach equilibrium, and a multiplicative rate reduction becomes more and more dramatic; in particular, long fat pipes are a poor match for TCP congestion control. See Chapter 4 for discussions of these facts as well as proposed solutions.

Regular loss: Since AIMD relies on loss as a congestion indicator, its inherent fluctuations require queues to grow (delay to increase) and lead to packet drops; this is even possible with ECN if many flows increase their rates at the same time.

Load-based charging: Because the sending rate depends, among other factors, on packet loss and the RTT, it is hard to properly monitor, trace or control the rate anywhere else in the network except at the node where the sender or receiver is located.

Applicability to interactive multimedia: The rate of an AIMD flow can underlie wild fluctuations; this can make TCP-like congestion control unsuitable for streaming media (see Sections 4.5 and 4.5.2).

In Chapter 4, we have seen that there are several proposals for protocols that correct some of these faults and therefore outperform TCP in one way or another. Still, significant enhancements are prevented by the TCP-friendliness requirement, which makes them downward compatible to technology that is outdated at its core and just was not designed for the environment of today. Given all these facts, it might be desirable to take a bold step forward and replace the stringent framework that is represented by the notion of TCP-friendly congestion control with a new and better one. For any such effort to be realistic, it should nevertheless be gradually deployable. Note that things become increasingly hypothetical at this point; still, I undertake an effort to sketch how this could be done:

- First of all, in order to justify its deployment, the framework should clearly be much better than TCP-friendly congestion control – a conforming mechanism should yield better QoS (less jitter, less loss, better link utilization, ..) than TCP, it should satisfy a well-defined measure of fairness (I already claimed that this would ideally be achieved by maximizing the respective utility functions of each sender), and there must be no doubts about its scalability and stability. All these things must be shown analytically and by extensive simulations and field trials.

- It should be possible to support tailored network usage as described in Section 6.3; to this end, the framework must be flexible enough to accommodate a large variety of mechanisms while still maintaining the properties above (stability, scalability etc.).

- Any conforming mechanism should be *slightly more aggressive* than TCP when competing with it. This, again, must be based on an analysis of how severe the impact of slightly more aggressive but still responsive mechanisms is, leading to a concise definition of 'slightly more aggressive'. Here, the idea is that users should be given an incentive to switch to a new scheme because it works slightly better than legacy mechanisms (and the old ones perform slightly worse as deployment of the new framework proceeds). It is obvious that this part of the design must be undertaken with special care in order to avoid degrading the service too severely (a significant part of the work should be devoted to finding out what 'degrading a service too severely' actually means).

- The complete design of such a new framework should be based upon *substantial* community (IETF) agreement; this might in fact be the most unrealistic part of the idea.

Eventually, since each user of the new framework would attain a certain local advantage from it, deployment could be expected to proceed quickly – which is dangerous if it is not designed very carefully. A lot of effort must therefore be put into proving that it remains stable under realistic Internet conditions.

6.2 Incentive issues

Congestion control is a very important function for maintaining the stability of the Internet; even a single sender that significantly diverges from the rules (i.e. sends at a high rate without responding to congestion) can impair a large number of TCP flows, thereby causing a form of congestion collapse (Floyd and Fall 1999). Still, today, congestion control is largely voluntary. Their operating system gives normal end users a choice between TCP and UDP – the former realizes congestion control, and the latter does not. TCP is more widely used, which is most probably because of the fact that it also provides reliability, which is a function that (i) is not easy to implement efficiently and (ii) is required by many applications. Since controlling the rate may only mean degraded throughput from the perspective of a single application, this situation provokes a number of questions:

1. What if we had a protocol that provides reliability without congestion control in our operating systems? Would we face a 'tragedy of the commons' (see Section 2.16) because selfish application programmers would have no incentive to use TCP?

2. Similarly, what prevents Internet users from changing, say, the TCP implementation in their operating system? This may just be impossible in the case of Microsoft Windows, but in an open-source operating system such as Linux it can easily be changed. Why is it that Linux users do not commonly patch their operating systems to make them act maliciously, as described in Section 3.5? The 'ACK division' attack, for instance, can be expected to yield a significant performance improvement,

as the majority of TCP stacks do not seem to implement appropriate byte counting (Section 4.1.1) (Medina et al. 2005).

Operating-system changes would not even have to be restricted to TCP: a patch that always sets ECT = 1 in UDP packets would misuse the technology, and probably lead to reduced loss in the presence of ECN-capable routers while not causing any harm to a single user of such an altered operating system.

3. It is widely known that applications that use UDP should realize congestion control, and they would ideally do so in a manner that is fair towards TCP. This is known to be difficult to implement, and the benefits for a single application are questionable – so what do UDP-based applications really do?

Answering the first two questions is really hard; I leave it to you to guess the answer to the first one. In the case of the second one, lack of information may be a reason – perhaps the people who would want to 'hack' their operating systems in order to speed things up simply do not know about these possibilities (should we then hope that they are not reading this book?). Another reason may be lack of incentives – if the effort appears to be greater than the gain, it is just not worth it. The effort for the ECN patch may be very small, but on the other hand, there are ECN deployment issues that may cause some people to hesitate regarding its usage (see Section 3.4.9). Finally, in order to answer the third question, we need to look at some measurements.

6.2.1 The congestion response of UDP-based applications

Using the isolated test bed shown in Figure 6.2, where five PCs are interconnected using Fast-Ethernet links (100 Mbps) and hubs, we carried out a number of simple measurements at the University of Innsbruck; specifically, we studied the congestion response of three streaming media applications, four VoIP tools, four video conferencing programs and four games. In one case, the same tool was used for VoIP and video conferencing; thus, the total

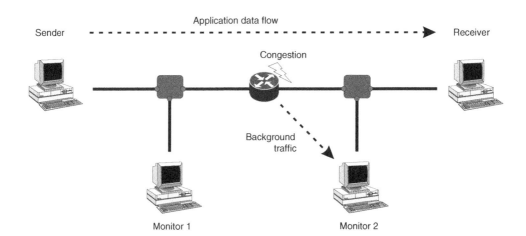

Figure 6.2 The test bed that was used for our measurements

number of applications is 14. The method of our measurements is always the same: two PCs running Windows 2000 act as sender and receiver (e.g. streaming server and client). A PC running Linux[3] with two network interfaces is used as a router. At the router interface that is connected to the PC labelled 'Monitor 2' in the figure, the available traffic rate is limited to by using the `tc` ('Traffic Control') Linux command and class-based queuing with only one class. We do not use token buckets because of their influence on traffic characteristics, for example, adverse effect on TCP (Huston 2000). The monitors, both running Linux, measure payload traffic *before* and *after* congestion, respectively. Loss is calculated as the difference between the bytes seen by Monitor 1 minus the bytes seen by Monitor 2. Background traffic is generated as a constant bit rate (UDP) data flow of 1000 byte packets using the `mgen` traffic generator[4]. It is sent from the router to Monitor 2, which means that it cannot cause collisions but can only lead to congestion in the queue of the router's outgoing network interface.

We generated three types of background traffic: *little*, *medium* and *much* traffic, which was scaled according to the type of software under study and always lasted a minute. Payload traffic generation depended on the type of application – for instance, for the streaming media tools, the same movie (a trailer of 'Matrix Reloaded') was encoded and played numerous times, while the games required replaying specific representative scenes (situations with lots of opponents, situations without opponents etc.) over and over again. All measurements lasted two minutes: 30 seconds without congestion, followed by a minute of congestion and 30 seconds without congestion again.

Our results varied wildly; some applications did not seem to respond to congestion at all, while others decreased their rate. A number of applications actually *increased* the rate in response to congestion. One can only guess what the reasons for such inappropriate behaviour may be – perhaps it is a way to compensate for loss via some means such as FEC. Some of the applications altered their rate by changing the packet size, while some others changed the actual sending rate (the spacing between packets). As an example, Figure 6.3 shows the sender rate and throughput of the popular streaming video tools 'Real Player', 'Windows Media Player' and 'Quicktime' under similar conditions ('medium' background traffic); TCP is included in the figure for comparison with the 'ideal' behaviour.

In these diagrams, a large gap between the two lines is a bad sign because it indicates loss. The figure shows how Real Player briefly increased its rate when congestion set in only to decrease it a little later (notably, it also decreased its packet size); it did not seem to recover when the congestion event was over. Windows Media Player reduced its rate, but it always seemed to send a little too much in our measurements, and Quicktime generally appeared to do the best job at adapting to the available bandwidth. Note that these results, like all others in this section, depend on the exact software version that was used and other factors; these details are not included here for the sake of brevity. It is *not* intended to judge the quality of these applications in this section – the idea is merely to provide an overview of the variety of congestion responses that real applications exhibit. For in-depth analyses of Real Player and Media Player, see (Mena and Heidemann 2000), (Wang et al. 2001), (Li et al. 2002) and (Nichols et al. 2004).

In our VoIP tests, 'Skype', 'ICQ' and 'Roger Wilco' did not respond to congestion – they sent at a steady albeit small rate. 'MSN' increased the size of its packets to

---

[3]RedHat 8.0, Linux kernel v2.4.18.
[4]http://mgen.pf.itd.nrl.navy.mil/mgen.html

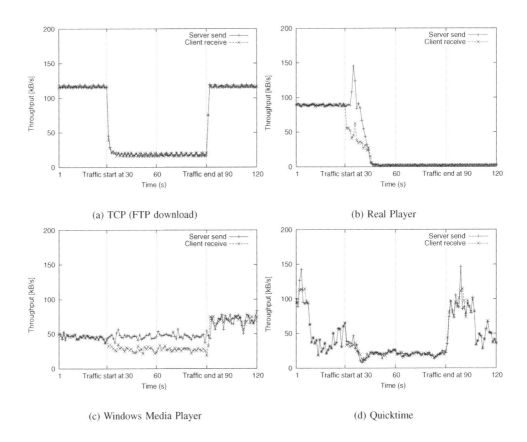

(a) TCP (FTP download)   (b) Real Player

(c) Windows Media Player   (d) Quicktime

Figure 6.3 Sender rate and throughput of streaming media tools

almost twice the original size in response to congestion, but did not react in any other way. This is only a single step reaction, that is, once the tool noticed congestion, it maintained its larger packet size no matter how much background traffic we used. Notably, some of these tools also detect and react to speech pauses, typically by decreasing the sending rate. Skype however maintained its rate during speech pauses by sending more but smaller packets. ICQ also did not react to congestion when we used its video conferencing capabilities; the same is true for 'AOL Messenger' and 'Microsoft Netmeeting', while 'iVisit' decreased both its rate and packet size in response to congestion. Finally, we tested four interactive multiplayer games: 'Jedi Knight 2' and 'Battle Field 1942' (first-person shooters), 'FIFA Football 2004' (a football simulation) and 'MotoGP 2' (a motorcycle racing game). Jedi Knight 2, which uses the well-known 'Quake 3' engine and may therefore represent a large number of other games, maintained its packet-sending rate but briefly used larger packets in response to congestion; as shown in Figure 6.4, FIFA Football 2004 also increased the size of its packets. The two other games did not react to congestion.

To conclude, only four out of the fourteen examined applications reduced their rate in response to congestion, and these four were sending a video stream. At least in two

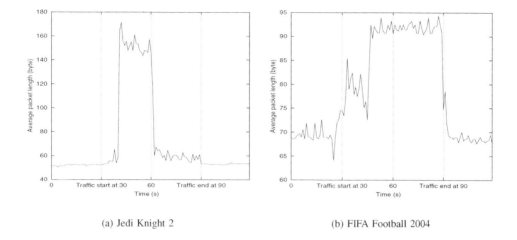

(a) Jedi Knight 2   (b) FIFA Football 2004

Figure 6.4 Average packet length of interactive multiplayer games

cases (Real Player and Windows Media Player), the way these applications reacted was questionable at best. One VoIP tool and two games increased the size of their packets and thereby increased the amount of traffic that they sent in the presence of congestion; all the other programs did not seem to react at all. I believe that this result can be taken as an indication that most applications do not contain congestion control functionality, and even the ones that do might not do the best job. What seems to be missing are *incentives* to do this correctly.

6.2.2 Will VoIP cause congestion collapse?

RFC 3714 (Floyd and Kempf 2004) discusses concerns regarding the fact that VoIP applications may not always properly react to congestion. As explained in the previous section, none of the four VoIP tools that we examined at the University of Innsbruck did. Theoretically, such unresponsive applications can severely harm TCP flows and lead to congestion collapse – this point was already made several times in this book, and the issue was stated quite clearly in (Floyd and Fall 1999), RFC 2914 (Floyd 2000) and RFC 2309 (Braden et al. 1998). The plea does not seem to help much; let us take a closer look at the reasons that might have led to this situation.

A single VoIP application does not generate a lot of traffic, and it is therefore not very likely to experience significant loss in today's overprovisioned networks. Or is it? RFC 3714 elaborates on the fact that overprovisioning the core does not eliminate congestion if access links are small; the document contains an example of a VoIP connection between a hotel room in Atlanta, Georgia, and Nairobi, Kenya, where the flow filled half of the capacity of the access link, and states that, for economic reasons, the growth rate of VoIP is likely to be greatest in developing countries where such access links are common. Such links may however not (yet?) be the common test bed of VoIP application developers, and I would assume that the same is true for the majority of users who download and utilize these tools. Actually, I know of at least two examples where this is true: I have a friend here in Innsbruck, Austria, who regularly uses Skype to talk to his girlfriend in China, and another friend uses the same tool to speak with his business partners in Miami, USA. Both are fascinated by the quality they obtain, and the recent success of Skype indicates that they are not alone.

The scenario that we are facing today is that VoIP applications work great for most customers even though they do not react to congestion. As long as they work well, customers will be pleased – and as long as they are, there will be no incentive for the authors of VoIP tools to implement congestion control. Loss is known to be critical for audio telephony – thus, in order to work well, it must not experience high loss rates. In my opinion, this is achieved by having individual peer-to-peer (endpoint-to-endpoint) applications transmit a low-rate data stream across an overprovisioned network, where the small access link problem simply does not occur. As the popularity of VoIP rises, we are facing a growing aggregate of unresponsive traffic in the network. The big picture appears to consist of three players, none of whom have a clear incentive to change anything:

1. *Application programmers* will not implement congestion control unless they have a reason to do so, that is, their customers complain (or switch to a different tool).

2. *End customers* will be happy when the application itself works fine and it does not significantly disturb other traffic. This is the case as long as the (constant) rate of a VoIP application clearly stays underneath the capacity of access links (because of core overprovisioning, such links are nowadays typically the bottleneck – but, as discussed in Section 2.1.1, this may change in the future).

3. *Service providers* could one day face links where the aggregate of unresponsive VoIP traffic causes significant harm to TCP flows. Assume that a service provider has such a problem – what should it do? Restraining the bandwidth of VoIP traffic or even blocking it will force customers to say 'ouch' (or rather 'good bye', which is worse). Since overprovisioning is currently quite a common strategy to solve anything, the ISP is likely to increase its link capacities.

We have already seen at the beginning of Chapter 2 where increasing link capacities to alleviate congestion from unresponsive traffic leads us. It seems that this vicious circle can only be stopped by giving at least one of the players an incentive to change the strategy. One such incentive could indeed come from Internet (and, in particular, VoIP) growth in developing countries – if a significant number of customers *do* have very small capacity access links, they *will* complain, and this could convince application developers to look for a solution. On the other hand, the small amount of traffic that is generated by VoIP tools may simply not be enough to cause severe problems, and even if it did, there is another control loop: when the telephone quality is poor, we hang up. Note that this is a difference between VoIP and streaming media, where some users may just keep an application running and do something else in the meantime if the quality is poor. Also, users might be willing to pay extra money to their ISPs for increasing the capacity when VoIP does not work anymore.[5]

---

[5]I would like to thank the members of the IRTF end-to-end interest mailing list who brought up these thoughts in the context of a related discussion.

In my opinion, what makes the VoIP problem especially interesting is the fact that it may serve as a role model regarding incentives and fully decentralized peer-to-peer style applications. Interactive multiplayer games, for instance, often use central servers, which will automatically lead to scalability problems (and the 'healthy' complaining customers) without congestion control – such problems have already been analysed in the literature (Feng et al. 2002a).

6.2.3 DCCP deployment considerations

Having discussed incentive issues with congestion control for VoIP, let us now focus on the question of DCCP deployment. I believe this to be a similarly intricate issue: one can easily argue that using, say, SCTP brings a benefit for certain special applications. That is a little more difficult for DCCP. From the perspective of a programmer who ponders whether to use the protocol, some potential immediate benefits of DCCP to the application are listed below:

- If it is planned to have thousands of users connect to a single server, it may work (i.e. scale) better because congestion control is properly implemented.

- TCP-based applications that are used at the same time may work better.

- The programmer's own application might experience a smaller loss ratio while maintaining reasonable throughput, that is, there are perhaps greater chances that most of what it sends really makes it to the other end, while its rate is almost as large as it can be under such circumstances. The fact that DCCP supports ECN might help here – in (Phelan 2004), this is mentioned as one of the motivating factors for using the protocol. On the other hand, it is questionable whether the mere fact that setting the ECN flag in a UDP-based flow is not enabled by the socket interface can prevent users from setting it in open-source operating systems such as Linux. This was already discussed at the beginning of Section 6.2 – maybe the reason for not doing so is the deployment problem of ECN itself? In this case, the possibility of using an ECN-enabled protocol may not seem so inviting.

Given that we might be talking about an application that would need to be updated to use the protocol and satisfies customers already, as well as the standard deployment problems with firewalls and so on, one may doubt that these arguments are convincing. The DCCP problem statement draft (Floyd et al. 2002) states:

> There has been substantial agreement (Braden et al. 1998; Floyd and Fall 1999) that in order to maintain the continued use of end-to-end congestion control, router mechanisms are needed to detect and penalize uncontrolled high-bandwidth flows in times of high congestion; these router mechanisms are colloquially known as 'penalty boxes'.

Now, let us assume that the only truly convincing argument to use DCCP would be the dramatically worse performance from UDP as the result of such penalty boxes. In this case, DCCP deployment can only happen once such boxes are widely used. An ISP will only have an incentive to install such a box if it yields a benefit – that is, if the financial loss from congestion problems with UDP traffic is greater than the loss from customers who switch to a different ISP because their UDP-based application does not work anymore. If a large majority of applications use UDP instead of DCCP, the latter loss may be quite significant. Thus, an ISP might have to wait for DCCP to be used by applications before installing penalty boxes, which in turn would motivate application designers to use DCCP – two parties waiting for each other, and what have we learned from history? Take a look at this paragraph about QoS deployment from RFC 2990 (Huston 2000):

> No network operator will make the significant investment in deployment and support of distinguished service infrastructure unless there is a set of clients and applications available to make immediate use of such facilities. Clients will not make the investment in enhanced services unless they see performance gains in applications that are designed to take advantage of such enhanced services. No application designer will attempt to integrate service quality features into the application unless there is a model of operation supported by widespread deployment that makes the additional investment in application complexity worthwhile and clients who are willing to purchase such applications. With all parts of the deployment scenario waiting for the others to move, widespread deployment of distinguished services may require some other external impetus.

Will we also need such other external impetus for DCCP, and what could it be?

6.2.4   Congestion control and QoS

All the incentive-related problems discussed so far have one common reason: implementing congestion control is expensive, and its direct or indirect benefit to whoever implements it is unclear. Thus, promoting a QoS oriented view of congestion control may be a reasonable (albeit modest) first step towards alleviating these issues. In other words, the question 'what is the benefit for a single well-behaved application?' should be a central one. In order to answer it, let us focus on the specific goal of congestion control mechanisms: *they try to use the available bandwidth as efficiently as possible*. One could add 'without interfering with others' to this statement, but this is the same as saying 'without exceeding it', which is pretty much the same as 'efficiently using it'. When you think about it, it really all seems to boil down to this single goal.

So what exactly does 'efficiently using the available bandwidth' mean? There are two sides to this. We have already seen the negative effects that loss can have on TCP in Chapter 3 – clearly, avoiding it is the most important goal. This is what was called *operating the network at the knee* in Section 2.2 and put in the form of a rule on Page 8: 'queues should generally be kept short'. Using as much of the available bandwidth as possible is clearly another goal – nobody wants a congestion control mechanism that always sends at a very low rate across a high-capacity path. This is underlined by the many research efforts towards a more-efficient TCP that we have discussed in Chapter 3. To summarize, we end up with two very simple goals, which, in order of preference, are as follows:

1. Keep queues short.

2. Utilize as much as possible of the available capacity.

The benefits of the first goal are low loss and low delay. The benefit of the second goal is high throughput – ideally, a congestion control mechanism would yield the highest possible throughput that does not lead to increased delay or loss.

At the 'Protocols for Fast Long-Distance Networks' workshop of 2005 in Lyon, France, a panel discussion revolved around the issue of benchmarking: there are many papers that somehow compare congestion control mechanisms with each other, but the community lacks a clearly defined set of rules for doing so. How exactly does one judge whether the performance of mechanism a is better than the performance of mechanism b? I think that someone should come up with a measure that includes (i) low loss and low delay and (ii) high capacity, and assigns a greater weight to the first two factors. Fulfilling these two goals is certainly attractive, and if a single, simple measure can be used to show that a 'proper' congestion control mechanism (as in TCP) does a better job at attaining these goals than the simplistic (or non-existing) mechanisms embedded in the UDP-based applications of today, this could yield a clear incentive to make use of a congestion control mechanism.

Still, the burden of having to implement it within the application certainly weighs on the negative side, and DCCP could help here – but DCCP might face the aforementioned deployment issues unless using congestion control is properly motivated. To summarize, I believe that we might need an integrated approach, where DCCP is provided and the benefits of using it are clearly stated using a simple measure as described above.

Looking at congestion control mechanisms from a QoS-centric perspective has another interesting facet: as discussed in (Welzl and Mühlhäuser 2003), such mechanisms could be seen as a central element for achieving fine-grain QoS in a scalable manner. This would be an alternative to IntServ over DiffServ as described in Section 5.3.5, and it could theoretically be realized with an architecture that comprises the following elements:

- A new service class would have to be defined; let us call it 'CC' service. The CC service would consist of flows that use a single specific congestion control mechanism only (or perhaps traffic that adheres to a fairness framework such as 'TCP-friendliness' or the one described in Section 6.1.2). Moreover, it would have to be protected from the adverse influence of other traffic, for example, by placing it in a special DiffServ class.

- Flows would have to be checked for conformance to the congestion control rules; this could be done by monitoring packets, tracking the sending rate and comparing it with a calculated rate in edge routers.

- Admission control would be needed in order to prevent unpredictable behaviour – there can be no guarantees if an arbitrary number of flows can enter and leave the network at any time. This would require some signalling, for example, a flow could negotiate its service with a 'bandwidth broker' that can also say 'not admitted at this time'. Again, this function could be placed at the very edges of the network.

This approach would scale, and the per-flow service quality attained with this approach could be calculated from environment conditions (e.g. number of flows present) and an analytic model of the congestion control mechanism or framework that is used. The throughput of TCP, for instance, can rather precisely be predicted with Equation 3.7 under such circumstances.

Note that the idea of combining congestion control with QoS is not a new one; similar considerations were made in (Katabi 2003) and (Harrison et al. 2002), and a more detailed sketch of one such architecture can be found in (Welzl 2005a).

6.3   Tailor-made congestion control

Nowadays, the TCP/IP transport layer does not offer much beyond TCP and UDP – but if we assume that the more-recent IETF developments will actually be deployed, we can expect Internet application programmers to face quite a choice of transport services within the next couple of years:

- Simple unreliable datagram delivery (UDP)

    - with or without delivery of erroneous data (UDP-Lite)

- Unreliable congestion-controlled datagram delivery (DCCP)

    - with or without delivery of erroneous data

    - with a choice of congestion control mechanisms

- Reliable congestion-controlled in-order delivery of

    - a consecutive data stream (TCP)

    - multiple data streams (SCTP)

- Reliable congestion-controlled unordered but potentially faster delivery of logical data chunks (SCTP)

This is only a very rough overview: each protocol has a set of features and parameters that can be tuned – for example, the size of an SCTP data chunk (or UDP or DCCP datagram) represents a trade-off between end-to-end delay and bandwidth utilization (small packets are delivered faster but have a larger per-packet header overhead than large packets). We have already discussed TCP parameter tuning in Section 4.3.3, and both SCTP and DCCP clearly have even more knobs that can be turned. Nowadays, the problem of the transport layer is its lack of flexibility. In the near future, this may be alleviated by DCCP and SCTP – but then, it is going to be increasingly difficult to figure out which transport protocol to use, how to use it and how to tune its parameters. The problem will not go away. Rather, it will turn into the issue of coping with transport layer complexity.

6.3.1   The Adaptation Layer

'Tailor-made Congestion Control' is a research project that just started at the University of Innsbruck. In this project, we intend to alleviate the difficulty of choosing and tuning a transport service by hiding transport layer details from applications; an underlying 'Adaptation Layer' could simply 'do its best' to fulfil application requirements by choosing from what is available in the TCP/IP stack. This method also makes the application independent

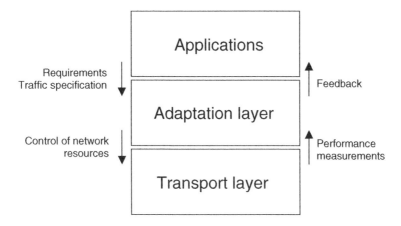

Figure 6.5 The Adaptation Layer

of the underlying network infrastructure – as soon as a new mechanism becomes available, the Adaptation Layer can use it and the application could automatically work better. The envisioned architecture is shown in Figure 6.5. As indicated by the arrows, an application specifies

- its network requirements (e.g. by providing appropriate weights to tune the trade-off between factors such as bandwidth, delay, jitter, rate smoothness and loss);

- the behaviour that it will show (e.g. by specifying whether it is 'greedy' (i.e. does it use all the resources it is given)).

On the basis of this information, the Adaptation Layer controls resources as follows:

- By choosing and tuning a mechanism from the stack – depending on the environment, this could mean choosing an appropriate congestion control mechanism and tuning its parameters via the DCCP protocol, making use of a protocol such as UDP-Lite, which is tailored for wireless environments, or even using an existing network QoS architecture.

- By performing additional functions such as buffering or simply choosing the ideal packet size. It is not intended to mandate strict layering underneath the Adaptation Layer; it could also directly negotiate services with interior network elements.

The Adaptation Layer needs to monitor the performance of the network. It provides QoS feedback to the application, which can then base its decisions upon this tailored high-level information rather than generic low-level performance measurements. As shown in Figure 6.6, it is yet another method to cope with congestion control state.

6.3.2 Implications

Introducing a layer above TCP/IP comes at the cost of service granularity – it will not be possible to specify an API that covers each and every possible service and at the same time

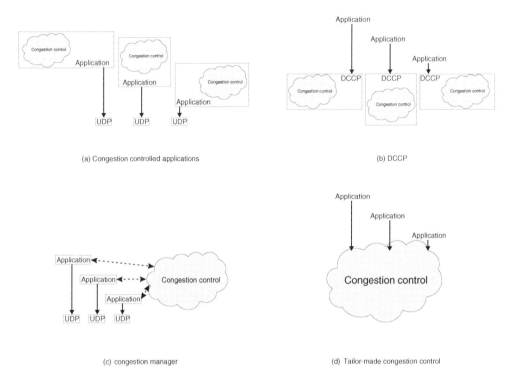

Figure 6.6 Four different ways to realize congestion control

ideally exploit all available mechanisms in any situation. On the positive side, the fact that applications could transparently use new services as they become available – much like SCTP-based TCP would provide multihoming without requiring to actually implement or even know about the feature at the application level – makes it possible to incrementally enhance implementations of the Adaptation Layer. This would automatically have a positive impact on the quality that is attained by users. For instance, if an ISP provides QoS to its users (who are running Adaptation Layer–enabled applications), there would be an immediate benefit; this is different from the situation of today, where someone would first have to write a QoS-enabled application that may not run in different environments. We believe that the Adaptation Layer could effectively break the vicious circle that is described in RFC 2990 (see Page 212) and thereby add a new dimension of competitiveness to the ISP market as well as the operating system vendor market.

The latter would be possible by realizing sophisticated mechanisms in the Adaptation Layer – this could even become a catalyst for related research, which would encompass the question whether it would be better to statically choose the right service or whether the Adaptation Layer should dynamically adapt to changing network characteristics. While it seems obvious that the latter variant could be more efficient, this brings about a number of new questions, for example, if the middleware switches between several TCP-friendly congestion control mechanisms depending on the state of the network, is the outcome still TCP-friendly? Since the Adaptation Layer could be aware of the network infrastructure, it could also make use of TCP-unfriendly but more-efficient transport protocols such as the ones described in Chapter 4. Would this be a feasible way to gradually deploy these mechanisms?

I have asked many questions and given few answers; while I am convinced that the Adaptation Layer would be quite beneficial, it could well be that only a small percentage of what I have described in this section is realistic. In our ongoing project, we work towards our bold goal in very small steps, and I doubt that we will ever realize everything that I have described here. On the other hand, as we approach the end of the chapter, I would like to remind you that being controversial was the original goal; I wanted to stir up some thoughts and provoke you to come up with your own ideas, and it would surprise me if all these bold claims and strange ideas did not have this effect on some readers. This chapter is an experiment of sorts, and as such it can fail. I did my best, and I sincerely hope that you found it interesting.

Appendix A

Teaching congestion control with tools

Teaching concepts of congestion control is not an easy task. On the one hand, it is always good to expose students to running systems and let them experience the 'look and feel' of real things; on the other, congestion control is all about traffic dynamics and effects that may be hard to see unless each student is provided with a test bed of her own. At the University of Innsbruck, we typically have lectures with 100 or more students, and accompanying practical courses with approximately 15 to 20 students per group. Every week, a set of problems must be solved by the students, and some of them are picked to present their results during a practical course. Compared to the other universities that I have seen, I would assume that our infrastructure is pretty good; we have numerous PC labs where every student has access to a fully equipped PC that runs Linux and Windows and is connected to the Internet. Still, carrying out real-life congestion control tests with these students would be impossible, and showing the effects of congestion control in an active real-life system during the lecture is a daunting task.

At first, having students carry out simulations appeared to solve the problem, but our experience was not very positive. Sadly, the only network simulator that seemed to make sense for us – ns – has quite a steep learning curve. Since we only wanted them to work on short and simple exercises that illustrate some important aspects of congestion control, students had to spend more than half of their time learning how to use the tool, while the goal was to have them spend almost all of the time with congestion control and nothing else. ns is not interactive, and the desired 'look and feel' exposure is just not there. In order to enhance this situation somewhat, two tools were developed at the University of Innsbruck. Simplicity was the main goal of this endeavour; the tools are small Java programs that only provide the minimum amount of features that is required, making them easy to use. They are presented in the following two sections along with hints on how they can be used in a classroom.

Network Congestion Control: Managing Internet Traffic   Michael Welzl
© 2005 John Wiley & Sons, Ltd

The tools can be obtained from the accompanying web site[1] of this book. You are encouraged to download, use and extend it at will; any feedback is welcome. Note that these tools were specifically designed for our environment, where we would like to show animations during the lecture and have groups of 15 or 20 students solve a problem in the classroom or as a homework (in which case it should be very simple and easy to present the outcome afterwards). If your setting allows for more flexibility (e.g. your groups of students are smaller, or they are doing projects that take longer), you may want to make use of more sophisticated tools such as the following:

tcpdump: This shows traffic going across an interface.[2]

ethereal: This does the same, but has a graphical user interface that can also be used to visualize tcpdump data.[3]

dummynet: This emulates a network and can be used to analyse the behaviour of an application under controlled conditions.[4]

NIST Net: This is like a dummynet, but runs under Linux and comes with a graphical user interface.[5]

A good overview of such tools can be found in (Hassan and Jain 2004).

A.1   CAVT

As we have seen in Section 2.5.1 of Chapter 2, vector diagrams are a very simple means of analysing the behaviour of congestion control mechanisms. At first sight, the case of two users sharing a single resource may appear to be somewhat unrealistic; on the other hand, while these diagrams may not suffice to prove that a mechanism works well, it is likely that a mechanism that does not work in this case is generally useless. Studying the dynamic behaviour of mechanisms with these diagrams is therefore worthwhile – but in scenarios that are slightly more complex than the ones depicted in Figure 2.4 (e.g. if one wants to study interactions between different mechanisms with homogeneous or even heterogeneous RTTs), imagining what the behaviour would look like or drawing a vector diagram on paper can become quite difficult.

This gap is filled by the *Congestion-Avoidance Visualization Tool (CAVT)*, a simple yet powerful program to visualize the behaviour of congestion control mechanisms. Essentially being a simulator that builds upon the well-known Chiu/Jain vector diagrams, the small Java application provides an interactive graphical user interface where the user can set a starting point and view the corresponding trajectory by clicking the mouse in the diagram. This way, it is easy to test a congestion control mechanism in a scenario where the RTTs are not equal: the rate of a sender simply needs to be updated after a certain number of time units. In the current version, precise traffic information is fed to the sources and no

---

[1]http://www.welzl.at/congestion
[2]http://www.tcpdump.org
[3]http://www.ethereal.com
[4]http://freshmeat.net/projects/dummynet
[5]http://www-x.antd.nist.gov/nistnet

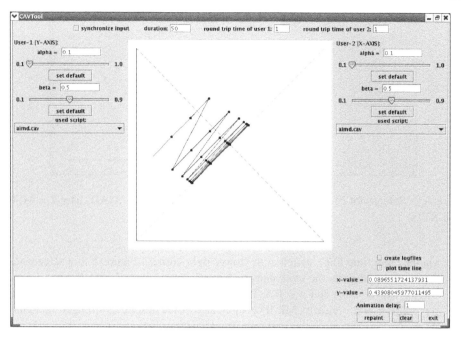

Figure A.1 Screenshot of CAVT showing an AIMD trajectory

calculations are done at routers. Also, there is no underlying queuing model; thus, assuming perfectly fluid behaviour, the state of the network during time instances t_{i-x}, t_i and t_{i+x} is seen by the source with RTT x at time instances t_i, t_{i+x} and t_{i+2x} respectively.

Figure A.1 is a screenshot of CAVT showing an AIMD trajectory; time is visualized by the line segments becoming brighter as they are drawn later in (simulated) time, but this effect is not noticeable with AIMD and equal RTTs because its convergence causes several lines to overlap. In addition to tuning individual parameters for each rate-allocation strategy, it is possible to individually change the RTTs and generate log files. Depending on the mechanism that was chosen, a varying number of parameter values with corresponding sliders is visible on each side of the drawing panel. Log files contain the trajectory seen in the GUI as well as the evolvement (time on the x-axis) of the rate of each user and the distance d of the current point p from the optimum o, which is simply the Euclidean distance:

$$d = \sqrt{(o_x - p_x)^2 + (o_y - p_y)^2} \qquad (A.1)$$

The white space in the lower left corner is used for status and error messages. CAVT has the following features:

- A facility to choose between various existing mechanisms such as AIMD, MIMD, AIAD and CADPC (see Section 4.6.4). These mechanisms were defined with a very simple script language, which can be used for further experimentation.

- An additional window that shows time on the x-axis plotted against the rate of each user as well as the distance from the optimum during operation (enabled or disabled

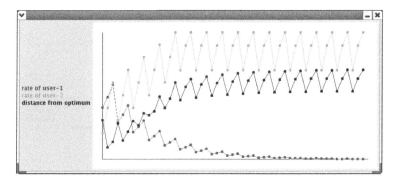

Figure A.2 Screenshot of the CAVT time line window (user 1 = AIAD, user 2 = MIMD, equal RTTs)

via the "plot time line" switch at the lower right corner). Figure A.2, a screenshot of this window, depicts the evolvement of a trajectory where AIAD competes against MIMD with $\alpha = 0.1$ and $\beta = 0.5$. The lowest line represents user 1 and the upper line represents user 2 – clearly, MIMD is more aggressive despite its stronger fluctuations. Also, the distance from the optimum (middle line) does not appear to converge to zero.

- Creation of log files that can be used with common plotting tools such as `xgraph` or `gnuplot` (enabled or disabled via the "create logfiles" switch at the lower right corner) – this is how Figures 2.4, 2.5 and 2.7 were generated.

- GUI control of the trajectory drawing speed in order to facilitate recognition of the evolvement over time.

The fact that CAVT can illustrate the behaviour of a mechanism with heterogeneous RTTs is particularly interesting, as it effectively provides researchers and students with a new abstraction level – a fluid-flow model with two users and heterogeneous RTTs. Table A.1 shows how this relates to other common abstraction levels for network analysis and design.

Table A.1 Common abstraction levels for network analysis and design

Level	Users	RTTs	Resources	Traffic model	Method
1	1	Equal	1	Fluid	Maths
2	2	Equal	1	Fluid	Vector diagrams
3	n	Equal	1	Fluid	Maths
4	**2**	**Different**	**1**	**Fluid**	**CAVT**
5	n	Different	1	Discrete	Simulation
6	n	Different	m	Discrete	Simulation
7	n	Different	m	Discrete	Real-life experiments

A.1.1 Writing scripts

The current version of CAVT can visualize arbitrary congestion control mechanisms as long as their decisions are based on traffic feedback only and do not use factors such as delay (as with TCP Vegas (Brakmo et al. 1994)); this covers all the TCP-friendly mechanisms that can be designed with the framework presented in (Bansal and Balakrishnan 2001) or even the "CYRF" framework in (Sastry and Lam 2002b). As mentioned earlier, a congestion control mechanism is specified using a simple script language. In this language, AIMD looks like this:

```
#define_param alpha range 0 to 1 default 0.1
#define_param beta range 0 to 1 default 0.5

    if(traffic < 1) {
        rate = rate + alpha;
    }
    else {
        rate = rate * beta;
    }
```

`alpha` and `beta` being the rate increase and decrease factors, respectively.

These parameters can be tuned using the GUI because the `define_param` command was used – this is a preprocessor directive that causes the GUI to display a slider and requires the user to define an upper and lower limit as well as a default value. The AIMD script given above was used to plot the trajectory in Figure A.1; the two sliders at the left- and right-hand side of the GUI therefore correspond with the `alpha` and `beta` parameters in the script. The CAVT script language defines two special variables:

1. `traffic` – the recently measured traffic

2. `rate` – the rate of the sender

In the present version, the only available control structures are the sequence and the `if` statement; loops are not available as they naturally occur every RTT. Variables, which are always of type "floating point", follow the standard naming convention found in languages such as Java or C and do not need to be declared (a variable is implicitly declared as it is assigned a value). In addition to the standard math operations, the following commonly known functions can be used: `abs`, `sin`, `cos`, `tan`, `exp`, `int`, `max`, `min`, `pow`, `sqrt` and `random`. If, for instance, we would like to assume that an AIMD sender cannot exceed the capacity because it is limited by the first link, we would need to change the script as follows:

```
#define_param alpha range 0 to 1 default 0.1
#define_param beta range 0 to 1 default 0.5
maxrate = 1.0; # rate can never exceed 100%

    if(traffic < 1) {
        rate = rate + alpha;
        rate = min(rate, maxrate);
    }
```

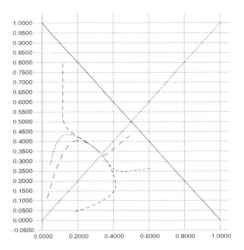

Figure A.3 Some CADPC trajectories

```
else {
    rate = rate * beta;
}
```

In the CAVT script language syntax, the code for CADPC (see Section 4.6.4) looks as follows:

```
#define_param a range 0.1 to 1 default 0.5

rate = rate * a * (1 - rate - traffic) + rate;
```

As explained in Section 4.6.4, CADPC does not directly converge to the optimum but to $n/(n + 1)$, where n is the number of flows in the system; this function rapidly converges to one (the normalized bottleneck capacity) as n increases. With two users, the point of convergence is 2/3, and this is evident in Figure A.3, which was generated with CAVT and *xgraph* (see Sections A.2.1 and A.2.3) using the above code.

CAVT does not contain a development environment; the script files can be generated using a standard text editor. Upon execution (scripts are not compiled but interpreted), errors are shown in the white area at the bottom of the screen. As soon as a file with the ending ".cav" is found in the CAVT home directory, it is made available in the drop down menu of each user (to the left and the right of the trajectory canvas). These graphical elements are shown in Figure A.1.

A.1.2 Teaching with CAVT

When teaching congestion control, there are two basic ways to make use of CAVT:

1. Use it as an animation tool during the lecture, which students can later download as a learning aid.

2. Use it to have students prepare exercises.

The first method (animation) worked fine in our case, although it is not quite clear how many students really used it as a learning aid. Test results indicate that the majority of them managed to understand what AIMD is all about. We also successfully used it for practical exercises by giving them the following homework:

1. Download CAVT and run it; make sure that you can answer the following questions and demonstrate the following tasks during the next meeting:

   - What do the x- and y-axis represent in the diagram, and what is the meaning of the diagonal lines? Which point in the diagram represents the ideal state, and why?

   - Assume that both users have equal parameters and RTTs. What is the graphical effect of a single 'Additive Increase/Additive Decrease' step in the diagram, and what is the graphical effect of a 'Multiplicative Increase/Multiplicative Decrease' step?

   - Assume that both users have equal parameters and RTTs. Show a trajectory for AIAD, MIMD and AIMD and explain why only AIMD converges to the optimum.

   - Write a .cav file that implements MIAD. Explain the result.

2. With AIMD, the rate is constantly fluctuating around the optimum. One may wonder whether it would be possible to minimize these fluctuations by dynamically adapting the increase parameter α and the decrease parameter β such that the average throughput is higher.

   Modify the 'aimd.cav' file as follows: the initial values $\alpha = 0.1$ and $\beta = 0.5$ should be constant. Add a parameter $0.1 \leq \gamma < 1$. When congestion occurs, α should be decreased by γ, and β should be increased by calculating $\beta_{new} = \gamma * (1 - \beta_{old})$.

   Test this mechanism with different values for γ and different RTTs. Explain the result; would this mechanism work in real life? Justify your answer!

This example was well received by our students; they appeared to quickly grasp the concept, and it seems that this exercise spurred interest in congestion control.

A.1.3 Internals

There are numerous ways to extend this software: one could imagine drawing additional lines that are parallel to the efficiency line in order to illustrate various levels of congestion or including special network behaviour (mechanisms that independently change the traffic measurements would, for instance, make it possible to study ATM ABR switch mechanisms). It would make sense to include a fluid-flow model of a queue in order to enable studying the effect of RTT estimation. Such a queue could, of course, be extended to support active queue management mechanisms such as RED. Another idea would be to support additional senders – for instance, it would be possible to have the user select two out of an arbitrary number of senders or even implement a three-dimensional extension in order to directly visualize the effect that increasing the number of users has on a mechanism.

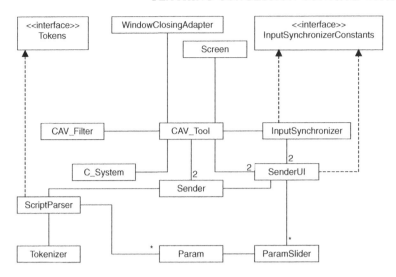

Figure A.4 Class structure of CAVT

It was suggested that integration of CAVT with the *ns* network simulator instead of building the individual elements from scratch be considered. However, we do not believe that this approach is fruitful: the strength of CAVT does not lie in its ability to visualize results from complex simulation scenarios (a function that is already provided by a wealth of visualization tools) but it lies in its interactive user interface that provides a user with a very direct method to 'play' with various congestion control strategies, which is helpful if one would like to gain more insight into their dynamics.

CAVT is an Open-source Java application. In order to facilitate its extension, it was designed in an object-oriented manner. Its simple class structure is shown as a UML diagram in Figure A.4; here, broken arrows represent the implementation of an interface and multiple associations are denoted by the symbols '2' (two classes) and '*' (an arbitrary number of classes). The diagram only contains classes that are not available in the standard JRE – dependencies with such classes are also not shown.

The central class is `CAV_Tool`; it manages all objects and graphical elements. There are two objects of type `Sender`, each of which has its own `ScriptParser` and `SenderUI`. When something is changed in the user interface of a sender (a `ParamSlider` is moved), the respective `Sender` communicates this change from the `SenderUI` to its `ScriptParser`, where a script file is interpreted using the current parameter settings (realized via the `Param` class). The `Tokenizer` is used to disassemble the script file into `Tokens` (which are defined in an interface) for parsing. The class `CAV_Filter` takes care of loading '.cav' files; the time-dependent trajectory plot is realized by the `C(anvas)_System`, which provides methods to add points and automatically scales the *x*-axis accordingly. The main window is represented by the `Screen` class. Finally, the `InputSynchronizer` function and the `InputSynchronizerConstraints` interface are related to a switch in the GUI, which is called *synchronize input*; if this switch is enabled, any parameter change for one user will automatically affect the parameters of the other user.

A.2 *ns*

The *ns* network simulator is a well-known open source tool. The facts that it is freely available online[6] and that the majority of mechanisms described in this book are or *can be* integrated in it – often, researchers who develop a new mechanism put their simulation code online, thereby enabling others to carry out analyses using the original code – makes it the ideal tool for studying congestion control. Specifically, there are implementations of all kinds of TCP variants (Tahoe, Reno, NewReno, SACK, Vegas), ECN, the TCP-friendly congestion control mechanisms TFRC, RAP and TEAR, XCP, CADPC/PTP and RED as well as several of its variants. The simulator, which is available for several Unix flavours and Windows (albeit not necessarily in the most recent version, as Unix is the main development platform), is sometimes called *ns* and sometimes called *ns-2* to account for the version number; we will use the terms synonymously.

ns is a so-called discrete event simulator: events are scheduled in advance by means of a simulation script that is written in 'OTcl' – an object-oriented variant of the Tcl programming language. Time consists of discrete steps; unless randomness is explicitly added, there are no fuzzy durations or time lags as there might be in real life because of effects at network nodes that are hard to predict. Thus, interfering factors such as process switching, the time to generate a packet (and reserve the required memory) and code execution timing variations are normally neglected. Architecturally, a split-programming model is used to separate time critical code (mechanisms such as low-level event processing and packet forwarding, which are written in C++) from simulation setup and dynamic configuration of network elements, which is done in OTcl.

Even a rough introduction to *ns* is beyond the scope of this book; please consult the main web page for further details. In what follows, basic knowledge of its fundamentals will be assumed.

A.2.1 Using *ns* for teaching: the problem

At first, *ns* is difficult to handle, which can mainly be attributed to lack of good documentation at the beginner level; it consists of some tutorials and the official '*ns* manual'. In our experience, the tutorials are too brief for providing students with all they need to know, while the manual mixes descriptions of how to *use* the simulator with descriptions of how to *change* it – which is reasonable for researchers who are mainly interested in integrating their own work, but inappropriate for a student who should merely attain some hands-on experience. At first, we tried to alleviate this problem by providing our own documentation in German language,[7] but that did not help much. This may be attributed to the fact that the documentation is not the only problem – the procedure for carrying out simulations is just exertive, and this may be an issue for students.

A typical simulation process is shown in Figure A.5: in this example, a user works with scripts in three programming languages, possibly applying minor changes to each. Also, several tools are used: a *text editor* is necessary to write code, simulations are run by executing *ns*, 'nam' output files can be visualized with the *network animator (nam)*

---

[6] http://www.isi.edu/nsnam/ns

[7] All these *ns* related things are available via the accompanying web site of this book, http://www.welzl.at/congestion

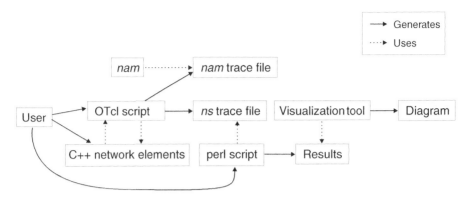

Figure A.5 A typical *ns* usage scenario. Reproduced by kind permission of Springer Science and Business Media

and results can easily be plotted with *xgraph*. Both *nam* and *xgraph* are included in the 'ns-allinone' download package that is available from the *ns* main web site.

The OTcl script describes the simulation scenario and uses *ns* objects, which, in turn, can call script functions in order to asynchronously return values. Two trace files are generated (as trace files can become huge and their generation can thus take long, in the case of large simulations, it is recommendable to only generate them as needed), one for direct visualization with *nam* and another one that, while being in an easy-to-parse format, requires further treatment before a result can be shown. Note that this is only one special way of using *ns*; while batch processing is an inherent property, there are numerous ways to trace network behaviour and generate data for further evaluation. For instance, things can be simplified somewhat at the cost of reduced flexibility by writing only the relevant data to a file directly from within a C++ network element. However, as explained in the following section, we do not believe that this should be done by students.

To summarize, while *nam* may be useful for showing animations in a lecture, we believe that there are two specific problems that make *ns* inappropriate as a tool for teaching exercises in a setting such as ours: (i) its cumbersome usage, and (ii) its non-interactive nature. In the following sections, I will explain how we tackled both of these problems.

A.2.2   Using *ns* for teaching: the solution

First of all, we came up with two rules for our exercises, which should make the task easier for students:

1. Never have students *change ns*; they should not have to worry about the C++ internals.

2. Never have students carry out more than one simulation in order to illustrate a single effect.

The second rule might require some further explanation. Typically, two kinds of analyses are carried out with *ns*: a single simulation can illustrate the dynamic behaviour of a certain mechanism – for example, how the congestion window of TCP fluctuates. Second, it is common to carry out a series of simulations with a single varying parameter, where each simulation yields only one result and the complete output can be used to illustrate certain dependencies. As an example, consider the following exercise that we gave to our students:

> Use *ns* to answer the question: 'Are multiple parallel FTP downloads faster than a single one?' Hint: study the throughput of 1 to 10 TCP flows that share a single bottleneck.

The goal is to show the total throughput as a function of the number of TCP flows. Therefore, the outcome should be a diagram that shows the varied parameter (the number of flows) on the x-axis and the total throughput on the y-axis; 10 simulations would need to be carried out. In order for the result to be meaningful, each individual simulation should last long enough for the mechanism to fully exhibit its typical behaviour; actually, not even determining the ideal (or minimum) duration is an easy task for students. Moreover, efficiently collecting the output from such a series of simulations can be quite troublesome (what if we had asked for the collective throughput of 1 to 100 TCP flows?).

There is an entirely different way to approach this problem. The result is not as visually appealing as the outcome of the above process, but it still suffices to illustrate the effect: students could do a single simulation, where they would prepare a scenario with 10 flows and simply not start all of them at the same time – that is, the first flow could be started right away, the second flow could be added after a couple of seconds and so on. This would lead to a diagram that shows how the average throughput grows over time, which we considered good enough for our purposes. This diagram might also show TCP fluctuations, which could be seen as an irrelevant side effect for this scenario, but we decided that we could live with this minor disadvantage of our method.

I have not yet explained how students are supposed to generate a diagram that shows throughput as a function of time. This can be accomplished by parsing trace files; we provided our students with a simple perl script (also converted to a Java program and a Windows executable) that does this job. This little program, which is called *throughput.pl*, writes two columns (time in the first column, throughput in the second column) to standard output; it takes the name of the script file, the protocol type and optional additional node number filters as input parameters. That is, if it is called as follows,

```
perl ./throughput.pl mytracefile.tr cbr 4
```

it will yield the throughput of constant bit rate traffic at node 4, and if it is called as follows,

```
perl ./throughput.pl mytracefile.tr cbr 4 1
```

it will only yield the throughput of constant bit rate traffic from node 1 at node 4. Redirecting the output to a file and simply calling 'xgraph file' will show the diagram.

Sticking with these restrictions and providing the perl script already made the task for students much easier. Still, one problem remained: the difficulty of writing OTcl script files.

A.2.3 NSBM

This brings us to the second tool that was developed in order to alleviate the task of teaching congestion control: *Network Simulation By Mouse (NSBM)*. Once again a small Java application, NSBM is quite easy to explain – it is a graphical user interface for creating

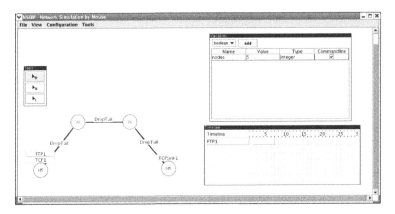

Figure A.6 A screenshot of NSBM

ns scripts. NSBM has all the standard features that one would expect, such as a menu that lets a user choose global parameter settings (including whether trace files should be generated, and what their names shall be) and a facility to save and load a scenario. Since the program is largely self-explanatory and comes with documentation, it will not be described in detail at this point; instead, its functions are explained by means of a simple example. Figure A.6 is a screenshot of NSBM. It shows the main elements of the tool:

Drawing canvas: The large white area (basically everything except for the three windows explained below) is used to draw the network topology, including network nodes and links as well as 'agents' and 'applications'.[8] The properties of any object can be changed via the right mouse button – clicking it on a node brings up a context window with a choice of 'properties', 'remove' or 'add agent'. In order to add an application, one would simply click the right mouse button on the corresponding agent.

Tools: This window is used to select one out of three mouse pointers that correspond with functions of the left mouse button; if the first (top) function is selected, nodes can be moved around using the left mouse button, and links will automatically follow. If the second selection is used, every click with the left mouse button will generate a network node, and the third selection lets the user draw lines from node to node (links that connect them). The function of the right mouse button always remains the same: it brings up a context window that enables the user to change properties of the object underneath the mouse pointer.

Timeline: This window shows all the 'applications' (application objects) that were generated in its left column. With the right mouse button, new 'run events' can be generated – the bar that ranges from time 1 to 7 means that the application 'FTP1' will start after 1 s and stop after 7 s. It can be moved, and its borders can be pulled to prolong or shorten the event.

---

[8]In *ns*, a TCP sender is an 'agent' that must be logically connected to a 'TCPSink' agent. In order to realize a greedy traffic source, an FTP 'application' must be stacked on top of the TCP agent. There are a large number of agents (different TCP variants etc.) and applications available.

Variables: Here, it is possible to define global variables for the simulation. When setting properties of an object, it is always possible to decide against specifying a constant and use a variable instead – this will confront the user with a list of variables to choose from (the ones that were defined in this window). In addition to the 'normal' properties that an *ns* object has, nodes have a property called *nodecount* in NSBM – this is the number of nodes that a single node represents, and it is indicated in the canvas by a number beside the letter 'N' in the circle. This provides for quite a bit of flexibility, as the nodecount can also depend on a variable. Moreover, the switch in the rightmost column of this window allows to turn any variable into a command line parameter.

The scenario shown in Figure A.6 is a simple dumbbell topology with *n* FTP senders (each of which will send data for 6 s) and receivers. After generating the code via 'Generate NS Code' in the 'File' menu, *ns* must be executed as follows:

```
ns code.tcl number_of_nodes
```

where 'code.tcl' is the name of the generated file and 'number_of_nodes' is the number of source and destination nodes. Figure A.7 is a screenshot of *nam* that shows the trace file that was obtained by calling the code with five nodes. Specifically, the name of the *nam* output file was set to 'out.nam' and the name of the generated file was set to 'code.tcl' in NSBM, the code was generated and the following commands were entered:

```
ns code.tcl 5
nam out.nam
```

Also, with 'out.tr' being the name of the regular *ns* trace file specified in NSBM, typing

```
perl ./throughput.pl out.tr tcp 6 > graph
xgraph graph
```

yielded Figure A.8, which is a screenshot of *xgraph*. As shown in Figure A.7, node number 6 is one end of the bottleneck; pressing the 'Play' button in *nam* clearly shows that it is the receiving end, as one can see packets flying from nodes 0 to 4 to nodes 7 to 11.

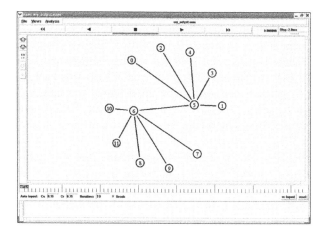

Figure A.7 A screenshot of *nam*

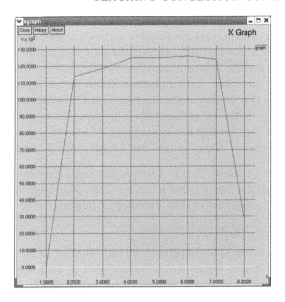

Figure A.8 A screenshot of *xgraph*

We have seen that NSBM and the 'throughput.pl' script make it very easy for students to create *ns* scenarios, the results of which can be plotted with *xgraph* and interactively studied with *nam*. All in all, it seems that the *ns* experience has turned from 80% batch processing (not 100% because of *nam*), to almost fully interactive usage. Moreover, it seems that using *ns* has become quite straightforward.

In fact, this is only half of the truth. First of all, it would be nice not to confront students with such a plethora of different tools, but only with one piece of software that provides all the necessary features, from scenario generation to simulation and visualization of results. Also, NSBM is not even flexible enough to cover all the scenarios that we would like our students to study. For instance, it is sometimes necessary to have them change parameters during a simulation – the simple 'run events' do not suffice here. Of course, it would have been possible to further extend NSBM, but we decided against it for the sake of simplicity. There is a fine line between adding features and keeping the software simple – and ease of use was the most-important goal for us. This being said, while NSBM only enables users to generate the agents and applications and handle the corresponding properties that we considered necessary by default, simple XML configuration files are used. By changing these files, support for new objects and their properties can easily be added to NSBM.

Our general approach is to have students go through the first couple of chapters of the famous 'Marc Greis Tutorial' that is available at the main *ns* project web page at the beginning; together with trying out the 'throughput.pl' perl script, this is their first exercise. Then, we show them NSBM and explain that we merely provide this tool as an additional aid, but that it will still be necessary for them to examine the OTcl code once in a while. Our experience with this method was very positive; it allowed students to focus on the congestion control-related details without being distracted by *ns* specific problems.

A.2.4 Example exercises

In addition to the 'multiple FTP' exercise on page 229, the following congestion control exercises worked well in our setting; we expect that they would generally be applicable as a homework for students who have access to our tools.

Phase effects: Download the simulator code for the 'congestion collapse' scenario in Section 2.2 from the accompanying web site of this book.[9] In addition to the update described in Section 2.2, change the capacity of the link from sender 1 to router 2 to 1 Mbps (thus, the links 0–1 and 1–2 now both have 1 Mbps). Run the simulation; you will notice the behaviour seen in Figure A.9. Explain this behaviour. Change the queue at router 2 from 'DropTail' to 'RED' – what changes, and why?

Window size: Simulate a simple network with two nodes and a link and show what happens when the maximum TCP window size is (i) too small, (ii) ideal, (iii) too large.

Basic TCP congestion control analysis: Simulate an FTP download with TCP Tahoe and analyse how TCP adapts its rate during the transmission. Use your simulation results to explain how the basic 'Tahoe' congestion control mechanism works: what are its phases and what happens during each phase?

Unresponsive senders: Explain why a single UDP flow can adversely affect the behaviour of many TCP flows. Illustrate this behaviour with an appropriate simulation scenario.

These are only some very simple examples; obviously, much more complicated scenarios can be studied with *ns*. In any case, I believe that NSBM, the 'throughput.pl' script and our two teaching rules on Page 228 can make your life – and the life of your students – easier.

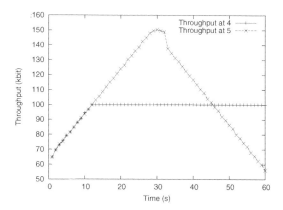

Figure A.9 The congestion collapse scenario with 1 Mbps from source 1 to router 2

---

[9]http://www.welzl.at/congestion

Appendix B

Related IETF work

B.1 Overview

The 'Internet Engineering Task Force (IETF)' is the standardization body of the Internet. It is open to everybody, and participation takes place online (in mailing lists) and in real life (there are regular meetings). The standardization process is too complex to be explained here in detail, and it has been explained in great depth in many other places (most importantly the IETF web site,[1] which is also your online starting point for everything related); instead of going into these boring details, this appendix provides you with brief 'what can I find where?' information.

The IETF consists of technical areas that encompass a number of working groups. As the name suggests, these working groups are where most of the work is done. You can always participate by visiting a group's web site (by following links from the main IETF web site), reading what it does, signing up to the mailing list and participating in the discussions. Usually, a working group develops something new – a new protocol, perhaps, or a set of security standards for dealing with attacks on existing protocols. Developments take the form of 'Internet-drafts', ASCII documents that have a limited lifespan of six months, much like some of the things in your refrigerator. By the way, this is also the case for working groups – thus, depending on when you read this, some of the working groups mentioned in this appendix may have already moved into the list of concluded working groups. Anyone can write Internet-drafts, and they are not reviewed for content before they are published – they are just a means to describe something that is being worked on in the IETF in a standardized form. While most Internet-drafts are developed by working groups, some are individual submissions.

After a thorough review, a request for publishing an Internet-draft as a 'Request for Comments (RFC)' can be made, which is actually a form of standard and not really a request for comments anymore. There are thousands of RFCs, and while these documents do not expire and will never be removed from the official list once they are published, they can have a different status. The three most important RFC categories are 'Standards

---

[1] http://www.ietf.org

Network Congestion Control: Managing Internet Traffic   Michael Welzl
© 2005 John Wiley & Sons, Ltd

Track', 'Informational' and 'Experimental'. While the latter two mean just what the words say (informational RFCs inform the community about something, while experimental RFCs describe something that you *can* implement if you are willing to participate in this experiment), there are several categories of standards track documents. 'Standard' documents are actual standards. 'Draft standards' are very close to becoming standards, and 'Proposed standards' are not so close.

At this point, it is worth noting that the IETF is quite conservative about changing existing protocols. This may be somewhat frustrating for researchers at first, but one must appreciate the role of the IETF: it is in charge of gradually enhancing TCP/IP technology while maintaining its stability. While the Internet has turned out to be immensely scalable, it has had its ups and downs here and there, and it *is* vulnerable, as pointed out in (Akella et al. 2002). Thus, the IETF is very cautious about introducing potentially disruptive technology or taking up new work unless it has been amply shown that it is worthwhile. This is also reflected in the status of documents: even RFC 2581, which specifies congestion control, only has 'Proposed standard' status. It is perhaps best to assume that some people will implement mechanisms that are specified in 'Proposed standard' RFCs, while there may not be so many volunteers when it comes to 'Experimental' RFCs. The IETF is not the place for preliminary research; it standardizes mechanisms once they are shown to be useful. The basis of decisions in the IETF is 'rough consensus and running code'.

In addition to the IETF, there is the IRTF[2] – the 'Internet Research Task Force'. As the name indicates, *this* is where research is discussed. There are also groups – 'Research groups' – but these are a bit different than IETF working groups: they meet on a much less regular basis, typically do not have milestones and working group items that must be taken care of, and some of them are not open to the public. One particular group of interest for congestion control researchers is the *End-to-End Research Group* (also called *end2end-interest*). You can find Van Jacobson's message that introduced Fast Retransmit and Fast Recovery in its mailing list archive. Theoretically, research groups can also cease their work, but this happens much less frequently than with IETF working groups, and the End-to-End Research Group, in particular, is quite old. Sometimes, a concrete piece of work evolves from IRTF discussions; one example of this process is the 'TCP Road map' document (Duke et al. 2004).

B.2   Working groups

Congestion control is a matter of the transport area. The following working groups (listed in no particular order) of this area deal or have dealt with congestion control in one way or another:[3]

tcpm: This is the place for minor updates of and extensions to TCP; at the time of writing, the F-RTO mechanism for spurious timeout detection (Sarolahti and Kojo 2004) and the TCP Roadmap document (Duke et al. 2004) were the documents this group was working on.

---

[2]http://www.irtf.org
[3]Note that this is a non-exhaustive list, and while these working groups were active at the time of writing, this might no longer be true when you read this. If you are still interested in further details, it might be helpful to know that, in addition to the working group list at http://www.ietf.org/html.charters/wg-dir.html, the IETF maintains a list of concluded working groups at http://www.ietf.org/html.charters/OLD/index.html

tsvwg: This is the place for transport area work that does not fit anywhere else – before TCPM came into being, it was also the group in charge of minor TCP extensions. Quite a number of the experimental enhancements of this protocol that were discussed in Chapter 4 were developed in this working group. Additionally, this group wrote some SCTP RFCs, RFC 3448 (Handley et al. 2003), which specifies TFRC, and the HighSpeed TCP specification document RFC 3649 (Floyd 2003).

dccp: This is where the Datagram Congestion Control Protocol (DCCP) is designed (see Section 4.5.2 in Chapter 4).

avt: AVT stands for Audio/Video Transport, and it is primarily concerned with RTP and its different payload formats. This rather old working group is relevant because multimedia applications that use RTP should execute some form of congestion control. In particular, the group is working on a realization of RTP-based TFRC (Gharai 2004), and it will ensure that RTP can efficiently be transported over Datagram Congestion Control Protocol DCCP.

rmt: The Reliable Multicast Transport (rmt) working group develops standards for one-to-many protocols; the TCP-friendly multicast congestion control protocol (Widmer and Handley 2004) is one of the items that this group is currently working on.

sigtran: This working group is concerned with PSTN signalling over IP networks; it developed SCTP.

pmtud: Path MTU Discovery as specified in RFC 1191 has some faults; in particular, it relies upon ICMP feedback from routers, which may never be generated (or permanently deleted by a firewall), thereby causing a problem called *Black Hole Detection*, that is, a source is fooled into believing that a destination is unreachable. In particular, this concerns TCP; some of its problems with PMTUD were documented in RFC 2923 (Lahey 2000). The goal of this working group is to enhance the mechanism and eliminate these errors. While Path MTU Discovery is not a congestion control mechanism, the scheme (and in particular, the new one that is being developed) is very closely connected to TCP and therefore, to some degree, also to its congestion control mechanisms.

rohc: This working group was mainly chartered to develop header compression schemes for RTP/UDP/IP, but it is also working on an enhanced header compression scheme that will support TCP options such as SACK or Timestamps. While the work does not directly relate to congestion control, having these options supported by a new header compression standard is likely to have some impact on congestion control in the Internet.

The following concluded working groups are noteworthy because of their congestion control-related efforts:

ecm: The 'Endpoint Congestion Management' working group wrote two RFCs: RFC 3124 (Balakrishnan and Seshan 2001), which specifies the congestion manager, and RFC 2914 (Floyd 2000), which specifies the 'Congestion Control Principles'.

pilc: PILC stands for 'Performance Implications of Link Characteristics'; this working group has written five RFCs about relationships between IETF protocols (especially TCP) and different link layers, including noisy and wireless links, as well as RFC 3135 (Border et al. 2001), which describes PEPs (see Section 4.3.1 in Chapter 4 for more details).

tcpimpl, tcplw, tcpsat: These groups have created some of the RFCs that make up the TCP standard and several RFCs that describe the experience with implementations or provide other related information to the community (e.g. RFC 2488 (Allman et al. 1999a), which describes how to use certain parts of the existing TCP specification for improving performance over satellites).

pcc: The 'Performance and Congestion Control' working group wrote RFC 1254 (Mankin and Ramakrishnan 1991), which is a survey of congestion control approaches in routers; among other things, it contains some interesting comments about ICMP Source Quench. This RFC was published in August 1991, and the working group concluded in the same year.

hostreq: This group, which also concluded in 1991, wrote RFC 1122 (Braden 1989), which was the first RFC to prescribe usage of the congestion control and RTO calculation mechanisms from (Jacobson 1988) in TCP.

B.3   Finding relevant documents

The intention of this appendix is to provide you with some guidance in your quest for congestion control-related information. The RFC index exceeded the number 4000 just a couple of days before I wrote this; simply browsing it is clearly no longer practical. The TCP standard has become so huge that even searching this list for the word 'TCP' is no longer a recommendable method for finding the RFCs that define this protocol. The best place to look for anything TCP related is the 'road map' document (Duke et al. 2004); in addition, Figure 3.10 on Page 88 puts the standards track RFCs that are somehow related to congestion control into a historic perspective.

For anything else, it may be good to start by checking the corresponding web pages of the working groups that deal or dealt with things that interest you. Finally, there are RFCs that are neither working group items nor a part of the TCP standard; most probably, for congestion control, the only important recent one is RFC 3714 (Floyd and Kempf 2004), 'IAB Concerns Regarding Congestion Control for Voice Traffic in the Internet'. If you really need to check the RFC list for exceptions such as this one, the ideal starting point is http://www.rfc-editor.org, but for most other relevant congestion control work, the sources above – combined with the links and bibliography of this book – will probably help you find anything you need.

Appendix C

List of abbreviations

ABC	Appropriate Byte Counting
ABE	Alternative Best-Effort
ABR	Available Bit Rate
ACK	Acknowledgement
AIAD	Additive Increase, Additive Decrease
AIMD	Additive Increase, Multiplicative Decrease
ALF	Application Layer Framing
AQM	Active Queue Management
ARQ	Automatic Repeat Request
ATM	Asynchronous Transfer Mode
AVQ	Adaptive Virtual Queue
AVT	Audio/Video Transport
BECN	Backwards ECN
BIC	Binary Increase TCP
CADPC	Congestion Avoidance with Distributed Proportional Control
CAPC	Congestion Avoidance with Proportional Control
CAVT	Congestion Avoidance Visualization Tool
CBR	Constant Bit Rate
CCID	Congestion Control ID
CE	Congestion Experienced
CHOKe	CHOose and Keep for responsive flows, CHOose and Kill for unresponsive flows
CLR	Current Limiting Receiver
CM	Congestion Manager
cwnd	Congestion Window
CWR	Congestion Window Reduced
CYRF	Choose Your Response Function
DCCP	Datagram Congestion Control Protocol
DDoS	Distributed Denial of Service
DRED	Dynamic-RED

Network Congestion Control: Managing Internet Traffic   Michael Welzl
© 2005 John Wiley & Sons, Ltd

DRS	Dynamic Right-Sizing
DupACK	Duplicate ACK
DWDM	Dense Wavelength Division Multiplexing
D-SACK	Duplicate-SACK
EC	Efficiency Controller
ECE	ECN-Echo
ECN	Explicit Congestion Notification
ECT	ECN Capable Transport
EPD	Early Packet Discard
EPRCA	Enhanced Proportional Rate Control Algorithm
ERICA	Explicit Rate Indication for Congestion Avoidance
ETEN	Explicit Transport Error Notification
EWMA	Exponentially Weighted Moving Average
FACK	Forward Acknowledgements
FC	Fairness Controller
FEC	Forward Error Correction/Forwarding Equivalence Class
FIFO	First In, First Out
FIN	Finalize
FLID-DL	Fair Layered Increase/Decrease with Dynamic Layering
FRED	Flow RED
GAIMD	General AIMD
GEOs	Geostationary Earth Orbit Satellite
HSTCP	HighSpeed TCP
ICMP	Internet Control Message Protocol
IETF	Internet Engineering Task Force
IIAD	Inverse Increase/Additive Decrease
IntServ	Integrated Services
IP	Internet Protocol
IPG	Inter-Packet Gap
IRTF	Internet Research Task Force
ISP	Internet Service Provider
IW	Initial Window
I-TCP	Indirect-TCP
LDA+	Loss-Delay Based Adjustment Algorithm
LDP	Label Distribution Protocol
LEOs	Low Earth Orbit Satellite
LFP	Long Fat Pipe
LSP	Label Switched Path
LSR	Label Switching Router
MIAD	Multiplicative Increase, Additive Decrease
MIMD	Multiplicative Increase, Multiplicative Decrease
MLDA	Multicast Loss-Delay Based Adjustment Algorithm
MPLS	Multi-Protocol Label Switching
MPλS	Multi-Protocol Lambda Switching
MSS	Maximum Segment Size
MTU	Maximum Transfer Unit

NACK	Negative Acknowledgement
NAT	Network Address Translator
NDP	Non-Data Packet
NE	Network Element
ns	Network Simulator
NSBM	Network Simulation By Mouse
NSIS	Next Steps in Signaling
OSU	Ohio State University (ATM ABR scheme)
PAWS	Protection Against Wrapped Sequence Numbers
PEP	Performance Enhancing Proxy
PGM	Pragmatic General Multicast
PHB	Per-Hop Behaviour
PH-RTO	Peak Hopper RTO
PLM	Packet Pair Receiver-driven Cumulative Layered Multicast
PMTUD	Path MTU Discovery
PPD	Partial Packet Discard
PSH	Push
PTP	Performance Transparency Protocol
QoS	Quality of Service
QS	Quick-start
RAP	Rate Adaptation Protocol
RED	Random Early Detection
RED-PD	RED with Preferential Dropping
REM	Random Early Marking
RFC	Request for Comments
RIO	RED with In/Out
RLC	Receiver-driven Layered Congestion Control
RLM	Receiver-driven Layered Multicast
RM	Resource Management (cells)
RMT	Reliable Multicast Transport
ROHC	Robust Header Compression
RR-TCP	Reordering-Robust TCP
RST	Reset
RSVP	Resource Reservation Protocol
RTCP	RTP Control Protocol
RTO	Retransmit Timeout
RTP	Real-time Transport Protocol
RTT	Round-trip Time
SACK	Selective Acknowledgements
SCTP	Stream Control Transmission Protocol
SFB	Stochastic Fair Blue
SFQ	Stochastic Fair Queuing
SIGTRAN	Signalling Transport
SRED	Stabilized RED
SRTT	Smoothed Round Trip Time
ssthresh	Slow start threshold

STCP	Scalable TCP
SWS	Silly Window Syndrome
SYN	Synchronize
TCB	TCP Control Block
TCP	Transmission Control Protocol
TCPM	TCP Maintenance and Minor Extensions
TEAR	TCP Emulation at Receivers
TFMCC	TCP-friendly Multicast Congestion Control
TFRC	TCP-friendly Rate Control
TSVWG	Transport Area Working Group
TTL	Time to live
UDP	User Datagram Protocol
URG	Urgent
VC	Virtual Circuit
VoIP	Voice over IP
WEBRC	Wave and Equation Based Rate Control
WRED	Weighted RED
XCP	eXpress Control Protocol

Bibliography

Akella A, Seshan S, Karp R, Shenker S and Papadimitriou C (2002) Selfish behavior and stability of the internet: a game-theoretic analysis of tcp. *SIGCOMM '02: Proceedings of the 2002 Conference on Applications, Technologies, Architectures, and Protocols for Computer Communications*, pp. 117–130. ACM Press.

Akyildiz IF, Morabito G and Palazzo S (2001) Tcp-peach: a new congestion control scheme for satellite ip networks. *IEEE/ACM Transactions on Networking* **9**(3), 307–321.

Akyildiz IF, Zhang X and Fang J (2002) Tcp-peach+: enhancement of tcp-peach for satellite ip networks. *IEEE Communications Letters* **6**(7), 303–305.

Allman M (2003) TCP Congestion Control with Appropriate Byte Counting (ABC) RFC 3465 (Experimental).

Allman M, Balakrishnan H and Floyd S (2001) Enhancing TCP's Loss Recovery Using Limited Transmit RFC 3042 (Proposed Standard).

Allman M, Dawkins S, Glover D, Griner J, Tran D, Henderson T, Heidemann J, Touch J, Kruse H, Ostermann S, Scott K and Semke J (2000) Ongoing TCP Research Related to Satellites RFC 2760 (Informational).

Allman M, Floyd S and Partridge C (2002) Increasing TCP's Initial Window RFC 3390 (Proposed Standard).

Allman M, Glover D and Sanchez L (1999a) Enhancing TCP Over Satellite Channels using Standard Mechanisms RFC 2488 (Best Current Practice).

Allman M and Paxson V (1999) On estimating end-to-end network path properties. *SIGCOMM '99: Proceedings of the Conference on Applications, Technologies, Architectures, and Protocols for Computer Communication*, pp. 263–274. ACM Press.

Allman M, Paxson V and Stevens W (1999b) TCP Congestion Control RFC 2581 (Proposed Standard). Updated by RFC 3390.

Andersson L, Doolan P, Feldman N, Fredette A and Thomas B (2001) LDP Specification RFC 3036 (Proposed Standard).

Appenzeller G, Keslassy I and McKeown N (2004) Sizing router buffers. *ACM SIGCOMM*, pp. 281–292. ACM Press.

Armitage G (2000) *Quality of Service in IP Networks: Foundations for a Multi-service Internet* Technology Series. MTR, Indianapolis, IN.

Ash J, Girish M, Gray E, Jamoussi B and Wright G (2002) Applicability Statement for CR-LDP RFC 3213 (Informational).

Athuraliya S, Low S, Li V and Yin Q (2001) Rem: active queue management. *IEEE Network* **15**(3), 48–53.

ATM Forum T (1999) Traffic Management Specification, Version 4.1. Technical Report AF-TM-0121.000, The ATM Forum.

Awduche D, Berger L, Gan D, Li T, Srinivasan V and Swallow G (2001) RSVP-TE: Extensions to RSVP for LSP Tunnels RFC 3209 (Proposed Standard). Updated by RFC 3936.

Awduche D, Chiu A, Elwalid A, Widjaja I and Xiao X (2002) Overview and Principles of Internet Traffic Engineering RFC 3272 (Informational).

Awduche D, Malcolm J, Agogbua J, O'Dell M and McManus J (1999) Requirements for Traffic Engineering Over MPLS RFC 2702 (Informational).

Aweya J, Ouellette M and Montuno DY (2001) A control theoretic approach to active queue management. *Computer Networks* **36**(2-3), 203–235.

Baker F (1995) Requirements for IP Version 4 Routers RFC 1812 (Proposed Standard). Updated by RFC 2644.

Bakre A and Badrinath BR (1995) I-tcp: indirect tcp for mobile hosts. *15th International Conference on Distributed Computing Systems (ICDCS)*, Vancouver, British Columbia, Canada.

Balakrishnan H, Padmanabhan V, Fairhurst G and Sooriyabandara M (2002) TCP Performance Implications of Network Path Asymmetry RFC 3449 (Best Current Practice).

Balakrishnan H, Rahul HS and Seshan S (1999) An integrated congestion management architecture for internet hosts. *SIGCOMM '99: Proceedings of the Conference on Applications, Technologies, Architectures, and Protocols for Computer Communication*, pp. 175–187. ACM Press.

Balakrishnan H and Seshan S (2001) The Congestion Manager RFC 3124 (Proposed Standard).

Balakrishnan H, Seshan S, Amir E and Katz R (1995) Improving tcp/ip performance over wireless networks. *1st ACM Conference on Mobile Communications and Networking (Mobicom)*, Berkeley, CA.

Balan RK, Lee BP, Kumar KRR, Jacob J, Seah WKG and Ananda AL (2001) TCP HACK: TCP header checksum option to improve performance over lossy links. *20th IEEE Conference on Computer Communications (INFOCOM)*, Anchorage, Alaska, USA.

Balinski M (2004) Fairness. die mathematik der gerechtigkeit. *Spektrum Der Wissenschaft.*

Bansal D and Balakrishnan H (2001) Binomial Congestion Control Algorithms. *IEEE INFOCOM 2001*, Anchorage, AK.

Barz C, Frank M, Martini P and Pilz M (2005) Receiver-based path capacity estimation for tcp. *Kommunikation in Verteilten Systemen (KiVS) – GI-Edition*, Kaiserslautern, Germany.

Bhandarkar S, Jain S and Reddy ALN (2005) Improving tcp performance in high bandwidth high rtt links using layered congestion control. *PFLDNet 2005 Workshop*, Lyon, France.

Blake S, Black D, Carlson M, Davies E, Wang Z and Weiss W (1998) An Architecture for Differentiated Service RFC 2475 (Informational). Updated by RFC 3260.

Blanton E and Allman M (2002) On making tcp more robust to packet reordering. *SIGCOMM Computer Communication Review* **32**(1), 20–30.

Blanton E and Allman M (2004) Using TCP Duplicate Selective Acknowledgement (DSACKs) and Stream Control Transmission Protocol (SCTP) Duplicate Transmission Sequence Numbers (TSNs) to Detect Spurious Retransmissions RFC 3708 (Experimental).

Blanton E, Allman M, Fall K and Wang L (2003) A Conservative Selective Acknowledgment (SACK)-based Loss Recovery Algorithm for TCP RFC 3517 (Proposed Standard).

Bless R, Nichols K and Wehrle K (2003) A Lower Effort Per-domain Behavior (PDB) for Differentiated Services RFC 3662 (Informational).

Bolot J (1993) End-to-end packet delay and loss behavior in the internet. *SIGCOMM '93: Conference Proceedings on Communications Architectures, Protocols and Applications*, pp. 289–298. ACM Press.

Border J, Kojo M, Griner J, Montenegro G and Shelby Z (2001) Performance Enhancing Proxies Intended to Mitigate Link-related Degradations RFC 3135 (Informational).

Bosau D (2005) Path tail emulation: an approach to enable end-to-end congestion control for split connections and performance enhancing proxies. In *Kommunikation in Verteilten Systemen KiVS 2005* (eds. Müller P, Gotzhein R and Schmitt JB), pp. 33–40. Gesellschaft für Informatik (GI), Kaiserslautern, Germany.

Braden R (1989) Requirements for Internet Hosts - Communication Layers RFC 1122 (Standard). Updated by RFC 1349.

Braden B, Clark D, Crowcroft J, Davie B, Deering S, Estrin D, Floyd S, Jacobson V, Minshall G, Partridge C, Peterson L, Ramakrishnan K, Shenker S, Wroclawski J and Zhang L (1998) Recommendations on Queue Management and Congestion Avoidance in the Internet RFC 2309 (Informational).

Braden R, Clark D and Shenker S (1994) Integrated Services in the Internet Architecture: An Overview RFC 1633 (Informational).

Braden R, Zhang L, Berson S, Herzog S and Jamin S (1997) Resource ReSerVation Protocol (RSVP) – Version 1 Functional Specification RFC 2205 (Proposed Standard). Updated by RFCs 2750, 3936.

Brakmo L, O'Malley S and Peterson L (1994) Tcp vegas: new techniques for congestion detection and avoidance. *ACM SIGCOMM*, pp. 24–35, London, UK.

Byers J, Frumin M, Horn G, Luby M, Mitzenmacher M, Roetter A and Shaver W (2000) Flid-dl: congestion control for layered multicast. *NGC'00*, Palo Alto, California, USA.

Campbell AT and Liao R (2001) A utility-based approach for quantitative adaptation in wireless packet networks. *Wireless Networks* **7**, 541–557.

Carlberg K, Gevros P and Crowcroft J (2001) Lower than best effort: a design and implementation. *SIGCOMM Computer Communication Review* **31**(suppl. 2), 244–265.

Carpenter B (1996) Architectural Principles of the Internet RFC 1958 (Informational). Updated by RFC 3439.

Carter RL and Crovella ME (1996) Measuring bottleneck link speed in packet switched networks. *Performance Evaluation* **27 and 28**, 297–318. Amsterdam, NY.

Cerf V, Dalal Y and Sunshine C (1974) Specification of Internet Transmission Control Program RFC 675.

Chait Y, Hollot CV, Misra V, Towsley D, Zhang H and Lui J (2002) Providing throughput differentiation for tcp flows using adaptive twocolor marking and two-level aqm. *IEEE Infocom*, New York.

Chen L, Nandan A, Yang G, Sanadidi MY and Gerla M (2004) Capprobe Based Passive Capacity Estimation. Technical Report TR040023, UCLA CSD.

Chen X, Mohapatra P and Chen H (2001) An admission control scheme for predictable server response time for web accesses. *WWW '01: Proceedings of The 10th International Conference On World Wide Web*, pp. 545–554. ACM Press, New York.

Chiu DM and Jain R (1989) Analysis of the increase/decrease algorithms for congestion avoidance in computer networks. *Computer Networks and ISDN Systems* **17**(1), 1–14.

Choe DH and Low SH (2004) Stabilized Vegas. In *Advances in Communication Control Networks, Lecture Notes in Control and Information Sciences*, (eds. Tarbouriech S, Abdallah C and Chiasson J) vol. 308 Springer Press.

Christin N, Grossklags J and Chuang J (2004) Near rationality and competitive equilibria in networked systems. *PINS '04: Proceedings of the ACM SIGCOMM Workshop on Practice and Theory of Incentives in Networked Systems*, pp. 213–219. ACM Press.

Clark D (1982) Window and Acknowledgement Strategy in TCP RFC 813.

Clark DD and Tennenhouse DL (1990) Architectural considerations for a new generation of protocols. *SIGCOMM '90: Proceedings of the ACM symposium on Communications Architectures & Protocols*, pp. 200–208. ACM Press.

Courcoubetis C and Weber R (2003) *Pricing Communication Networks*. John Wiley & Sons.

Crovella ME and Bestavros A (1997) Self-similarity in world wide web traffic: evidence and possible causes. *IEEE/ACM Transactions on Networking* **5**(6), 835–846.

Crowcroft J, Hand S, Mortier R, Roscoe T and Warfield A (2003) Qos's downfall: at the bottom, or not at all! *Proceedings of the ACM SIGCOMM Workshop on Revisiting IP QoS*, pp. 109–114. ACM Press.

Crowcroft J and Oechslin P (1998) Differentiated end-to-end internet services using a weighted proportional fair sharing tcp. *SIGCOMM Computer Communication Review* **28**(3), 53–69.

Dawkins S, Montenegro G, Kojo M and Magret V (2001a) End-to-end Performance Implications of Slow Links RFC 3150 (Best Current Practice).

Dawkins S, Montenegro G, Kojo M, Magret V and Vaidya N (2001b) End-to-end Performance Implications of Links with Errors RFC 3155 (Best Current Practice).

Deepak B, Hari Balakrishnan SF and Shenker S (2001) Dynamic behavior of slowly-responsive congestion control algorithms. *ACM SIGCOMM 2001*, San Diego, CA.

Diederich J, Lohmar T, Zitterbart M and Keller R (2000) Traffic phase effects with RED and constant bit rate UDP-based traffic. *QofIS*, pp. 233-244. Berlin, Germany.

Dovrolis C, Ramanathan P and Moore D (2001) What do packet dispersion techniques measure? *IEEE INFOCOM*, pp. 905–914, Anchorage, Alaska, USA.

Downey AB (1999) Using pathchar to estimate internet link characteristics. *SIGCOMM '99: Proceedings of the Conference on Applications, Technologies, Architectures, and Protocols for Computer Communication*, pp. 241–250. ACM Press.

Drewes M and Haid O (2000) *Tiroler Küche*. Tyrolia Verlaganstalt.

Duke M, Braden R, Eddy W and Blanton E (2004) A roadmap for Tcp Specification Documents, Internet-draft Draft-ietf-tcpm-tcp-roadmap-00, Work in Progress.

Durresi A, Sridharan M, Liu C, Goyal M and Jain R (2001) Traffic management using multilevel explicit congestion notification. *SCI 2001*, Orlando Florida. Special session 'ABR to the Internet'.

Eggert L, Heidemann J and Touch J (2000) Effects of ensemble-tcp. *SIGCOMM Computer Communication Review* **30**(1), 15–29.

Ekström H and Ludwig R (2004) The peak-hopper: a new end-to-end retransmission timer for reliable unicast transport. *IEEE Infocom*, Hong Kong, China.

Ely D, Spring N, Wetherall D, Savage S and Anderson T (2001) Robust congestion signaling. *Proceedings of International Conference on Network Protocols*, Riverside, CA.

Fairhurst G and Wood L (2002) Advice to Link Designers on Link Automatic Repeat ReQuest (ARQ) RFC 3366 (Best Current Practice).

Fall K and Floyd S (1996) Simulation-based comparisons of tahoe, reno, and SACK TCP. *Computer Communication Review* **26**(3), 5–21.

Feamster N, Bansal D and Balakrishnan H (2001) On the interactions between layered quality adaptation and congestion control for streaming video *11th International Packet Video Workshop*, Kyongju, Korea.

Feng W, Chang F, Feng W and Walpole J (2002a) Provisioning on-line games: a traffic analysis of a busy counter-strike server. *IMW '02: Proceedings of the 2nd ACM SIGCOMM Workshop on Internet Measurement*, pp. 151–156. ACM Press.

Feng W, Gardner MK, Fisk ME and Weigle EH (2003) Automatic flow-control adaptation for enhancing network performance in computational grids. *Journal of Grid Computing* **1**(1), 63–74.

Feng W, Kandlur D, Saha D and Shin K (1999) A self-configuring red gateway. *Conference on Computer Communications (IEEE Infocom)*, New York.

Feng W, Shin KG, Kandlur DD and Saha D (2002b) The blue active queue management algorithms. *IEEE/ACM Transactions on Networking* **10**(4), 513–528.

Fielding R, Gettys J, Mogul J, Frystyk H, Masinter L, Leach P and Berners-Lee T (1999) Hypertext Transfer Protocol – HTTP/1.1 RFC 2616 (Draft Standard). Updated by RFC 2817.

Floyd S (1994) TCP and Successive Fast Retransmits. ICIR, Available from http://www.icir.org/floyd/notes.html.

Floyd S (2000) Congestion Control Principles RFC 2914 (Best Current Practice).

Floyd S (2002) General Architectural and Policy Considerations RFC 3426 (Informational).

Floyd S (2003) HighSpeed TCP for Large Congestion Windows RFC 3649 (Experimental).

Floyd S (2004) Limited Slow-start for TCP with Large Congestion Windows RFC 3742 (Experimental).

Floyd S and Fall K (1999) Promoting the use of end-to-end congestion control in the internet. *IEEE/ACM Transactions on Networking* **7**(4), 458–472.

Floyd S, Gummadi R and Shenker S (2001) Adaptive RED: an algorithm for increasing the robustness of RED's active queue management. Available from http://www.icir.org/floyd/papers.html.

Floyd S, Handley M and Kohler E (2002) Problem Statement for dccp, Internet-draft Draft-ietf-dccp-problem-00, Work in Progress.

Floyd S, Handley M, Padhye J and Widmer J (2000a) Equation-based congestion control for unicast applications. *ACM SIGCOMM*, pp. 43–56. ACM Press.

Floyd S and Henderson T (1999) The NewReno Modification to TCP's Fast Recovery Algorithm RFC 2582 (Experimental). Obsoleted by RFC 3782.

Floyd S, Henderson T and Gurtov A (2004) The NewReno Modification to TCP's Fast Recovery Algorithm RFC 3782 (Proposed Standard).

Floyd S and Jacobson V (1991) Traffic phase effects in packet-switched gateways. *SIGCOMM Computer Communication Review* **21**(2), 26–42.

Floyd S and Jacobson V (1993) Random early detection gateways for congestion avoidance. *IEEE/ACM Transactions on Networking* **1**(4), 397–413.

Floyd S, Jacobson V, Liu CG, McCanne S and Zhang L (1997) A reliable multicast framework for light-weight sessions and application level framing. *IEEE/ACM Transactions on Networking* **5**(6), 784–803.

Floyd S and Kempf J (2004) IAB Concerns Regarding Congestion Control for Voice Traffic in the Internet RFC (3714) (Informational).

Floyd S and Kohler E (2005) Tcp Friendly Rate Control (tfrc) For Voice: Voip Variant and Faster Restart. Internet-draft Draft-ietf-dccp-tfrc-voip-01, Work in Progress.

Floyd S, Mahdavi J, Mathis M and Podolsky M (2000b) An Extension to the Selective Acknowledgement (SACK) Option for TCP RFC 2883 (Proposed Standard).

Fomenkov M, Keys K, Moore D and Claffy K (2004) Longitudinal study of internet traffic in 1998-2003. *WISICT '04: Proceedings of the Winter International Symposium on Information and Communication Technologies*, pp. 1–6. Trinity College Dublin.

Gemmell J, Montgomery T, Tony Speakman T, Bhaskar N and Crowcroft J (2003) The pgm reliable multicast protocol. *IEEE Network* **17**(1), 16–22.

Gerla M and Kleinrock L (1980) Flow control: a comparative survey. *IEEE Transactions on Communications* **28**(4), 553–574.

Gevros P, Risso F and Kirstein P (1999) Analysis of a method for differential tcp service. *IEEE GLOBECOM Symposium on Global Internet*, Rio De Janeiro, Brazil.

Gharai L (2004) Rtp Profile for Tcp Friendly Rate Control. Internet-draft draft-ietf-avt-tfrc-profile-03, Work in progress.

Gibbens R and Kelly F (1999) Distributed connection acceptance control for a connectionless network. In *Teletraffic Engineering in a Competitive World*, (eds. Key P and Smith D), ITC16, pp. 941–952. Elsevier, Amsterdam, NY.

Grieco LA and Mascolo S (2002) Tcp westwood and easy red to improve fairness in high-speed networks. *PIHSN '02: Proceedings of the 7th IFIP/IEEE International Workshop on Protocols for High Speed Networks*, pp. 130–146. Springer-Verlag.

Grieco LA and Mascolo S (2004) Performance evaluation and comparison of westwood+, new reno, and vegas tcp congestion control. *SIGCOMM Computer Communication Review* **34**(2), 25–38.

Grossglauser M and Tse DNC (1999) A framework for robust measurement-based admission control. *IEEE/ACM Transactions on Networking* **7**(3), 293–309.

Grossman D (2002) New Terminology and Clarifications for Diffserv RFC 3260 (Informational).

Group AVTW, Schulzrinne H, Casner S, Frederick R and Jacobson V (1996) RTP: A Transport Protocol for Real-time Applications RFC 1889 (Proposed Standard). Obsoleted by RFC 3550.

Gruber G (1994) *Langfristige Abhängigkeiten, Pseudozyklen und Selbstähnlichkeit in ökonomischen Zeitreihen*. Schriftenreihe 'Forschungsergebnisse der Wirtschaftsuniversität Wien, Vienna, Austria.

Gu Y and Grossman RL (2005) Optimizing udp-based protocol implementations. *PFLDNet 2005 Workshop*, Lyon, France.

Guo L, Crovella M and Matta I (2001) How does tcp generate pseudo-self-similarity? *Proceedings of the Ninth International Symposium in Modeling, Analysis and Simulation of Computer and Telecommunication Systems (MASCOTS'01)*, p. 215. IEEE Computer Society.

Guo L and Matta I (2001) The war between mice and elephants. *The 9th IEEE International Conference on Network Protocols (ICNP)*, Riverside, CA.

Gurtov A and Floyd S (2004) Modeling wireless links for transport protocols. *SIGCOMM Computer Communication Review* **34**(2), 85–96.

Gurtov A and Ludwig R (2003) Responding to spurious timeouts in tcp. *IEEE Infocom*, San Francisco, CA.

Handley M, Floyd S, Padhye J and Widmer J (2003) TCP Friendly Rate Control (TFRC): Protocol Specification RFC 3448 (Proposed Standard).

Handley M, Floyd S, Whetten B, Kermode R, Vicisano L and Luby M (2000a) The Reliable Multicast Design Space for Bulk Data Transfer RFC 2887 (Informational).

Handley M, Padhye J and Floyd S (2000b) TCP Congestion Window Validation RFC 2861 (Experimental).

Hardin G (1968) The tragedy of the commons. *Science* **162**, 1243–1248.

Harrison D, Kalyanaraman S and Ramakrishnan S (2002) Congestion control as a building block for qos. *SIGCOMM Computer Communication Review* **32**(1), 71.

Hasegawa G, Kurata K and Murata M (2000) Analysis and improvement of fairness between tcp reno and vegas for deployment of tcp vegas to the internet. *IEEE International Conference on Network Protocols (ICNP)*, pp. 177–186, Osaka, Japan.

Hassan M and Jain R (2004) *High Performance TCP/IP Networking*. Pearson Education International.

He G, Gao Y, Hou J and Park K (2002) A case for exploiting self-similarity in internet traffic in tcp congestion control. *IEEE International Conference on Network Protocols*, pp. 34–43, Paris, France.

Heinanen J and Guerin R (1999a) A Single Rate Three Color Marker RFC 2697 (Informational).

Heinanen J and Guerin R (1999b) A Two Rate Three Color Marker RFC 2698 (Informational).

Henderson T, Crowcroft J and Bhatti S (2001) Congestion pricing. Paying your way in communication networks. *IEEE Internet Computing* **5**(5), 85–89.

Heying Z, Baohong L and Wenhua D (2003) Design of a robust active queue management algorithm based on feedback compensation. *SIGCOMM '03: Proceedings of the 2003 Conference on Applications, Technologies, Architectures, and Protocols for Computer Communications*, pp. 277–285. ACM Press.

Hoe J (1995) *Start-up dynamics of TCP's congestion control and avoidance schemes.* Master's thesis, Massachusetts Institute of Technology, Department of Electrical Engineering and Computer Science.

Hoe JC (1996) Improving the start-up behavior of a congestion control scheme for tcp. *SIGCOMM '96: Conference Proceedings on Applications, Technologies, Architectures, and Protocols for Computer Communications*, pp. 270–280. ACM Press.

Hollot C, Misra V, Towsley D and Gong W (2001) On designing improved controllers for aqm routers supporting tcp flows. *INFOCOM 2001. Twentieth Annual Joint Conference of the IEEE Computer and Communications Societies*, vol. 3, pp. 1726–1734, Anchorage, Alaska, USA.

Huitema C (2000) *Routing in the Internet.* second edn. P T R Prentice-Hall, Englewood Cliffs, NJ 07632.

Hurley P, Le Boudec J, Thiran P and Kara M (2001) Abe: providing a low-delay service within best effort. *IEEE Network* **15**(3), 60–69.

Huston G (2000) Next Steps for the IP QoS Architecture RFC 2990 (Informational).

Inamura H, Montenegro G, Ludwig R, Gurtov A and Khafizov F (2003) TCP over Second (2.5G) and Third (3G) Generation Wireless Networks RFC 3481 (Best Current Practice).

ISO (1994) Is 7498 OSI(Open System Interconnection Model) Standard.

Jacobs S and Eleftheriadis A (1997) Real-time dynamic rate shaping and control for internet video applications. *Workshop on Multimedia Signal Processing*, pp. 23–25, Siena, Italy.

Jacobson V (1988) Congestion avoidance and control. *Symposium Proceedings on Communications Architectures and Protocols*, pp. 314–329. ACM Press.

Jacobson V (1990) Compressing TCP/IP headers for low-speed serial links RFC 1144 (Proposed Standard).

Jacobson V (1998) Notes on Using Red for Queue Management and Congestion Avoidance, Viewgraphs from Talk at Nanog 13, Available From ftp://ftp.ee.lbl.gov/talks/vj-nanog-red.pdf.

Jacobson V and Braden R (1988) TCP extensions for long-delay paths RFC 1072. Obsoleted by RFCs 1323, 2018.

Jacobson V, Braden R and Borman D (1992) TCP Extensions for High Performance RFC 1323 (Proposed Standard).

Jain A and Floyd S (2005) Quick-start for Tcp and Ip. internet-draft draft-amit-quick-start-04, Work in progress.

Jain R (1986) A timeout based congestion control scheme for window flow- controlled networks. *IEEE Journal on Selected Areas in Communications* **SAC-4**(7), 1162–1167. Reprinted in Partridge C (Ed.) *Innovations in Internetworking*, 289–295, Artech House, Norwood, MA 1988.

Jain R (1989) A delay based approach for congestion avoidance in interconnected heterogeneous computer networks. *Computer Communications Review, ACM SIGCOMM* pp. 56–71.

Jain R (1990) Congestion control in computer networks: trends and issues. *IEEE Network*, **4(3)**, 24–30.

Jain R, Chiu D and Hawe W (1984) A Quantitative Measure of Fairness and Discrimination for Resource Allocation in Shared Computer Systems. DEC Research Report TR-301, Digital Equipment Corporation, Maynard, MA.

Jain R and Ramakrishnan K (1988) Congestion avoidance in computer networks with A connectionless network layer: concepts, goals, and methodology. In *Computer Networking Symposium: Proceedings*, April 11–13, 1988 (ed. IEEE), pp. 134–143. IEEE Computer Society Press, Sheraton National

Hotel, Washington, DC area, Silver Spring, MD 20910. IEEE catalog number 88CH2547-8. Computer Society order number 835.

Jamin S, Shenker S and Danzig P (1997) Comparison of measurement-based admission control algorithms for controlled-load service. *IEEE Infocom 1997*, vol. 3, pp. 973–980, Kobe.

Jamoussi B, Andersson L, Callon R, Dantu R, Wu L, Doolan P, Worster T, Feldman N, Fredette A, Girish M, Gray E, Heinanen J, Kilty T and Malis A (2002) Constraint-based LSP Setup using LDP RFC 3212 (Proposed Standard). Updated by RFC 3468.

Jiang H and Dovrolis C (2004) The effect of flow capacities on the burstiness of aggregated traffic. *Passive and Active Measurements (PAM) conference*, Antibes Juan-les-Pins, France.

Jin C, Wei DX and Low SH (2004) Fast tcp: motivation, architecture, algorithms, performance *IEEE Infocom*, Hong Kong.

Johari R and Tan DKH (2001) End-to-end congestion control for the internet: delays and stability. *IEEE/ACM Transactions on Networking* **9**(6), 818–832.

Kalyanaraman S, Jain R, Fahmy S, Goyal R and Kim SC (1996) Performance and buffering requirements of internet protocols over ATM ABR and UBR services. *IEEE Communications Magazine* **36**(6), 152–157.

Kapoor R, Chen L, Lao L, Gerla M and Sanadidi MY (2004) Capprobe: a simple and accurate capacity estimation technique. *SIGCOMM '04: Proceedings of the 2004 Conference on Applications, Technologies, Architectures, and Protocols for Computer Communications*, pp. 67–78. ACM Press.

Karn P, Bormann C, Fairhurst G, Grossman D, Ludwig R, Mahdavi J, Montenegro G, Touch J and Wood L (2004) Advice for Internet Subnetwork Designers RFC 3819 (Best Current Practice).

Karn P and Partridge C (1987) Round trip time estimation. *ACM SIGCOMM-87*, Stowe, Vermont, USA.

Karn P and Partridge C (1995) Improving round-trip time estimates in reliable transport protocols. *SIGCOMM Computer Communication Review* **25**(1), 66–74.

Karsten M (2000) *QoS Signalling and Charging in a Multi-service Internet using RSVP*. PhD thesis, Darmstadt University of Technology.

Katabi D (2003) *Decoupling Congestion Control from the Bandwidth Allocation Policy and its Application to High Bandwidth-delay Product Networks*. PhD thesis, MIT.

Katabi D and Blake C (2002) A Note on the Stability Requirements of Adaptive Virtual Queue. Technical Report MIT-LCS-TM-626, MIT.

Katabi D, Handley M and Rohrs C (2002) Congestion control for high bandwidth-delay product networks. *SIGCOMM '02: Proceedings of the 2002 Conference on Applications, Technologies, Architectures, and Protocols for Computer Communications*, pp. 89–102. ACM Press.

Katz D (1997) IP Router Alert Option RFC 2113 (Proposed Standard).

Kelly F (1997) Charging and rate control for elastic traffic. *European Transactions on Telecommunications* **8**, 33–37.

Kelly FP, Maulloo A and Tan D (1998) Rate control in communication networks: shadow prices, proportional fairness and stability. *Journal of the Operational Research Society* **49**, 237–252.

Kelly T (2001) An ecn probe-based connection acceptance control. *SIGCOMM Computer Communication Review* **31**(3), 14–25.

Kelly T (2003) Scalable tcp: improving performance in highspeed wide area networks. *SIGCOMM Computer Communication Review* **33**(2), 83–91.

Kent S and Atkinson R (1998) Security Architecture for the Internet Protocol RFC 2401 (Proposed Standard). Updated by RFC 3168.

Keshav S (1991a) *Congestion Control in Computer Networks* PhD thesis UC Berkeley. published as UC Berkeley TR-654.

Keshav S (1991b) A control-theoretic approach to flow control *ACM SIGCOMM*, pp. 3–15. ACM Press.

Kohler E (2004) Datagram Congestion Control Protocol Mobility and Multihoming, Internet-draft Draft-kohler-dccp-mobility-00, Work in Progress.

Kohler E, Handley M and Floyd S (2005) Datagram Congestion Control Protocol (dccp), Internet-draft draft-ietf-dccp-spec-11, Work in Progress.

Krishnan R, Sterbenz J, Eddy W, Partridge C and Allman M (2004) Explicit transport error notification (eten) for error-prone wireless and satellite networks. *Computer Networks* **46**(3), 343–362.

Kumar A, Hedge M, Anand SVR, Bindu BN, Thirumurthy D and Kherani AA (2000) Nonintrusive tcp connection admission control for bandwidth management of an internet access link. *IEEE Communications Magazine* **38**(5), 160–167.

Kunniyur S and Srikant R (2001) Analysis and design of an adaptive virtual queue (avq) algorithm for active queue management. *SIGCOMM '01: Proceedings of the (2001) Conference on Applications, Technologies, Architectures, and Protocols for Computer Communications*, pp. 123–134. ACM Press.

Kurose JF and Ross KW (2004) *Computer Networking: A Top-down Approach Featuring the Internet*. Pearson Benjamin Cummings.

Laatu V, Harju J and Loula P (2003) Evaluating performance among different tcp flows in a differentiated services enabled network. *ICT 2003*, Tahiti, Papeete – French Polynesia.

Ladha S, Baucke S, Ludwig R and Amer PD (2004) On making sctp robust to spurious retransmissions. *SIGCOMM Computer Communication Review* **34**(2), 123–135.

Lahey K (2000) TCP Problems with Path MTU Discovery RFC 2923 (Informational).

Lai K and Baker M (2000) Measuring link bandwidths using a deterministic model of packet delay. *ACM SIGCOMM*, pp. 43–56. ACM Press.

Larzon LA, Degermark M, Pink S, Jonsson LE and Fairhurst G (2004) The Lightweight User Datagram Protocol (UDP-Lite) RFC 3828 (Proposed Standard).

Le Boudec J (2001) Rate Adaptation, Congestion Control and Fairness: A Tutorial.

Lee S, Seok S, Lee S and Kang C (2001) Study of tcp and udp flows in a differentiated services network using two markers system. *MMNS '01: Proceedings of the 4th IFIP/IEEE International Conference on Management of Multimedia Networks and Services*, pp. 198–203. Springer-Verlag.

Legout A and Biersack EW (2000) Plm: fast convergence for cumulative layered multicast transmission schemes. *SIGMETRICS '00: Proceedings of the 2000 ACM SIGMETRICS International Conference on Measurement and Modeling of Computer Systems*, pp. 13–22. ACM Press.

Legout A and Biersack E (2002) Revisiting the fair queuing paradigm for end-to-end congestion control. *IEEE Network* **16**(5), 38–46.

Leland WE, Taqqu MS, Willinger W and Wilson DV (1993) On the self-similar nature of ethernet traffic. *SIGCOMM '93: Conference Proceedings on Communications Architectures, Protocols and Applications*, pp. 183–193. ACM Press.

Li B and Liu J (2003) Multirate video multicast over the internet: an overview. *IEEE Network* **17**(1), 24–29.

Li M, Claypool M and Kinicki B (2002) Media player versus real player – a comparison of network turbulence. *ACM SIGCOMM Internet Measurement Workshop*, Marseille, France.

Lin D and Morris R (1997) Dynamics of random early detection. *SIGCOMM '97: Proceedings of the ACM SIGCOMM '97 Conference on Applications, Technologies, Architectures, and Protocols for Computer Communication*, pp. 127–137. ACM Press.

Low SH, Paganini F, Wang J, Adlakha S and Doyle JC (2002) Dynamics of tcp/red and a scalable control *IEEE Infocom*, New York City, New York, USA.

Luby M and Goyal V (2004) Wave and Equation Based Rate Control (WEBRC) Building Block RFC 3738 (Experimental).

Luby M, Goyal VK, Skaria S and Horn GB (2002) Wave and equation based rate control using multicast round trip time. *SIGCOMM '02: Proceedings of the 2002 Conference on Applications, Technologies, Architectures, and Protocols for Computer Communications*, pp. 191–204. ACM Press.

Ludwig R and Gurtov A (2005) The Eifel Response Algorithm for TCP RFC 4015(Proposed Standard).

Ludwig R and Katz RH (2000) The eifel algorithm: making tcp robust against spurious retransmissions. *SIGCOMM Computer Communication Review* **30**(1), 30–36.

Ludwig R and Meyer M (2003) The Eifel Detection Algorithm for TCP RFC 3522 (Experimental).

Luenberger DG (1979) *Introduction to Dynamic Systems*. John Wiley & Sons.

MacKie-Mason JK and Varian HR (1993) Pricing the internet. *Public Access to the Internet*, JFK School of Government, Harvard University, May 26–27, 1993, p. 37.

Mahajan R, Floyd S and Wetherall D (2001) Controlling high-bandwidth flows at the congested router. *9th International Conference on Network Protocols (ICNP)*, Riverside, California, USA.

Makridakis SG, Wheelwright SC and Hyndman RJ (1997) *Forecasting: Methods and Applications*. John Wiley & Sons.

Manimaran G and Mohapatra P (2003) Multicasting: an enabling technology. *IEEE Network* **17**(1), 6–7.

Mankin A and Ramakrishnan K (1991) Gateway Congestion Control Survey RFC 1254 (Informational).

Mankin A, Romanow A, Bradner S and Paxson V (1998) IETF Criteria for Evaluating Reliable Multicast Transport and Application Protocols RFC 2357 (Informational).

Mascolo S and Racanelli G (2005) Testing tcp westwood+ over transatlantic links at 10 gigabit/second rate. *Protocols for Fast Long-distance Networks (PFLDnet) Workshop*, Lyon, France.

Massoulie L (2002) Stability of distributed congestion control with heterogeneous feedback delays. *IEEE Transactions on Automatic Control* **47**(6), 895–902.

Mathis M and Mahdavi J (1996) Forward acknowledgement: refining tcp congestion control. *SIGCOMM '96: Conference Proceedings on Applications, Technologies, Architectures, and Protocols for Computer Communications*, pp. 281–291. ACM Press.

Mathis M, Mahdavi J, Floyd S and Romanow A (1996) TCP Selective Acknowledgement Options RFC 2018 (Proposed Standard).

Matrawy A and Lambadaris I (2003) A survey of congestion control schemes for multicast video applications. *IEEE Communications Surveys & Tutorials* **5**(2), 22–31.

McCanne S, Jacobson V and Vetterli M (1996) Receiver-driven layered multicast. *SIGCOMM '96: Conference Proceedings on Applications, Technologies, Architectures, and Protocols for Computer Communications*, pp. 117–130. ACM Press.

McCloghrie K and Rose M (1991) Management Information Base for Network Management of TCP/IP-based internets:MIB-II RFC 1213 (Standard). Updated by RFCs 2011, 2012, 2013.

Medina A, Allman M and Floyd S (2004) Measuring the Evolution of Transport Protocols in the Internet. Technical report, ICIR. To appear in ACM Computer Communication Review (CCR), April 2005.

Mena A and Heidemann J (2000) An empirical study of real audio traffic. *IEEE Infocom*, pp. 101–110, Tel-Aviv, Israel.

Metz C (1999) Tcp over satellite...the final frontier. *IEEE Internet Computing* **3**(1), 76–80.

Mills D (1983) Internet Delay Experiments RFC 889.

Mogul J, Brakmo L, Lowell DE, Subhraveti D and Moore J (2004) Unveiling the transport. *SIGCOMM Computer Communication Review* **34**(1), 99–106.

Mogul J and Deering S (1990) Path MTU Discovery RFC 1191 (Draft Standard).

Moore D, Voelker GM and Savage S (2001) Inferring internet denial-of-service activity *USENIX Security Symposium*, Washington DC.

Mortier R, Pratt I, Clark C and Crosby S (2000) Implicit admission control. *IEEE Journal on Selected Areas in Communications Special Issue: QoS in the Internet* **18**(12), 2629–2639.

Mullin J, Smallwood L, Watson A and Wilson GM (2001) New techniques for assessing audio and video quality in real-time interactive communication. *IHM-HCI Conference*, Lille, France.

Nagle J (1984) Congestion Control in IP/TCP Internetworks RFC 896.

Nichols K, Blake S, Baker F and Black D (1998) Definition of the Differentiated Services Field (DS Field) in the IPv4 and IPv6 Headers RFC 2474 (Proposed Standard). Updated by RFCs 3168, 3260.

Nichols J, Claypool M, Kinicki R and Li M (2004) Measurements of the congestion responsiveness of windows streaming media. *ACM International Workshop on Network and Operating Systems Support for Digital Audio and Video (NOSSDAV)*, Cork, Ireland.

Ott TJ and Aggarwal N (1997) Tcp over atm: abr or ubr? *SIGMETRICS '97: Proceedings of the 1997 ACM SIGMETRICS International Conference on Measurement and Modeling of Computer Systems*, pp. 52–63. ACM Press.

Ott T, Lakshman T and Wong L (1999) Sred: stabilized red. *INFOCOM '99. Eighteenth Annual Joint Conference of the IEEE Computer and Communications Societies*, vol. 3, pp. 1346–1355, New York City, New York, USA.

Padhye J, Firoiu V, Towsley D and Kurose J (1998) Modeling tcp throughput: a simple model and its empirical validation. *SIGCOMM '98: Proceedings of the ACM SIGCOMM '98 Conference on Applications, Technologies, Architectures, and Protocols for Computer Communication*, pp. 303–314. ACM Press.

Padhye J and Floyd S (2001) On inferring tcp behavior. *SIGCOMM '01: Proceedings of the 2001 Conference on Applications, Technologies, Architectures, and Protocols for Computer Communications*, pp. 287–298. ACM Press.

Paganini F, Wang Z, Low SH and Doyle JC (2003) A new tcp/aqm for stable operation in fast networks. *IEEE Infocom*, San Francisco, CA.

Pan R, Prabhakar B and Psounis K (2000) Choke - a stateless active queue management scheme for approximating fair bandwidth allocation. *INFOCOM 2000. Nineteenth Annual Joint Conference of the IEEE Computer and Communications Societies*, vol. 2, pp. 942–951, Tel-Aviv, Israel.

Park K and Willinger W (2000) *Self-similar Network Traffic and Performance Evaluation*. John Wiley & Sons.

Paxson V (1999) End-to-end internet packet dynamics. *IEEE/ACM Transactions on Networking* **7**(3), 277–292.

Paxson V and Allman M (2000) Computing TCP's Retransmission Timer RFC 2988 (Proposed Standard).

Paxson V, Allman M, Dawson S, Fenner W, Griner J, Heavens I, Lahey K, Semke J and Volz B (1999) Known TCP Implementation Problems RFC 2525 (Informational).

Paxson V and Floyd S (1995) Wide area traffic: the failure of poisson modelling. *IEEE/ACM Transactions on Networking* **3**(3), 226.

Perkins C (2002) IP Mobility Support for IPv4 RFC 3344 (Proposed Standard).

Peterson LL and Davie BS (2003) *Computer Networks: A Systems Approach, 3e*. Morgan Kaufmann Publishers.

Phelan T (2004) Datagram Congestion Control Protocol (dccp) User Guide, Internet-draft Draft-ietf-dccp-user-guide-02, Work in Progress.

Postel J (1981a) Internet Protocol RFC 791 (Standard). Updated by RFC 1349.

Postel J (1981b) Transmission Control Protocol RFC 793 (Standard). Updated by RFC 3168.

Ramakrishnan K and Floyd S (1999) A Proposal to Add Explicit Congestion Notification (ECN) to IP RFC 2481 (Experimental). Obsoleted by RFC 3168.

Ramakrishnan K, Floyd S and Black D (2001) The Addition of Explicit Congestion Notification (ECN) to IP RFC 3168 (Proposed Standard).

Rejaie R, Handley M and Estrin D (1999a) Quality adaptation for congestion controlled video playback over the internet. *SIGCOMM '99: Proceedings of the Conference on Applications, Technologies, Architectures, and Protocols for Computer Communication*, pp. 189–200. ACM Press.

Rejaie R, Handley M and Estrin D (1999b) Rap: an end-to-end rate-based congestion control mechanism for realtime streams in the internet. *IEEE Infocom*, New York City, New York, USA.

Rhee I, Ozdemir V and Yi Y (2000) Tear: Tcp emulation at receivers - flow control for multimedia streaming. NCSU Department of Computer Science. Available from http://www.csc.ncsu.edu/faculty/rhee/export/tear_page/

Rhee I and Xu L (2005) Cubic: a new tcp-friendly high-speed tcp variant. *Protocols for Fast Long-distance Networks (PFLDnet) Workshop*, Lyon, France.

Rizzo L (2000) Pgmcc: a tcp-friendly single-rate multicast congestion control scheme. *SIGCOMM '00: Proceedings of the Conference on Applications, Technologies, Architectures, and Protocols for Computer Communication*, pp. 17–28. ACM Press.

Romanow A and Floyd S (1994) Dynamics of tcp traffic over atm networks. *SIGCOMM '94: Proceedings of the Conference on Communications Architectures, Protocols and Applications*, pp. 79–88. ACM Press.

Saltzer JH, Reed DP and Clark DD (1984) End-to-end arguments in system design. *ACM Transactions on Computer Systems* **2**(4), 277–288.

Saltzer JH, Reed DP and Clark DD (1998) Active networking and end-to-end arguments. *IEEENET* **12**(3), 69–71.

Samaraweera N and Fairhurst G (1998) High speed internet access using satellite-based dvb networks. *International Network Conference (INC'98)*, Plymouth.

Sarolahti P and Kojo M (2004) Forward Rto-recovery (f-rto): An Algorithm for Detecting Spurious Retransmission Timeouts with Tcp and Sctp, Internet-draft draft-ietf-tcpm-frto-02, Work in progress.

Sarolahti P, Kojo M and Raatikainen K (2003) F-rto: an enhanced recovery algorithm for tcp retransmission timeouts. *SIGCOMM Computer Communication Review* **33**(2), 51–63.

Sastry N and Lam SS (2002a) Cyrf: A Framework for Window-based Unicast Congestion Control. Technical Report TR-02-09, Department of Computer Sciences, The University of Texas at Austin.

Sastry N and Lam SS (2002b) A theory of window-based unicast congestion control. *IEEE ICNP 2002*, Paris, France.

Savage S, Cardwell N, Wetherall D and Anderson T (1999) Tcp congestion control with a misbehaving receiver. *SIGCOMM Computer Communication Review* **29**(5), 71–78.

Savorić M, Karl H and Wolisz A (2003) The TCP control block interdependence in fixed networks — new performance results. *Computer Communications Magazine* **26**(4), 367–376.

Schulzrinne H, Casner S, Frederick R and Jacobson V (2003) RTP: A Transport Protocol for Real-time Applications RFC 3550 (Standard).

Semke J, Mahdavi J and Mathis M (1998) Automatic tcp buffer tuning. *ACM SIGCOMM*, Vancouver, British Columbia, Canada.

Shenker S (1994) Making greed work in networks: a game-theoretic analysis of switch service disciplines. *SIGCOMM '94: Proceedings of the Conference on Communications Architectures, Protocols and Applications*, pp. 47–57. ACM Press.

Shenker S (1995) Fundamental design issues for the future internet. *IEEE Journal on Selected Areas in Communications* **13**(7), 1176–1188.

Shorten RN and Leith DJ (2004) H-tcp: tcp for high-speed and long-distance networks. *PFLDnet Workshop*, Argonne, Illinois USA.

Simpson W (1995) IP in IP Tunneling RFC 1853 (Informational).

Sisalem D and Schulzrinne H (1996) Switch mechanisms for the abr service: a comparison study. *Proceedings of Telecommunication Distribution Parallelism (TDP)*, La Londes Les Maures, France.

Sisalem D and Wolisz A (2000a) Lda+: a tcp-friendly adaptation scheme for multimedia communication. In *IEEE International Conference on Multimedia and Expo - ICME*, pp. 1619–1622, New York City, New York, USA.

Sisalem D and Wolisz A (2000b) Mlda: a tcp-friendly congestion control framework for heterogeneous multicast environments. *8th International Workshop on Quality of Service (IWQoS 2000)*, pp. 65–74, Pittsburgh, PA.

Speakman T, Crowcroft J, Gemmell J, Farinacci D, Lin S, Leshchiner D, Luby M, Montgomery T, Rizzo L, Tweedly A, Bhaskar N, Edmonstone R, Sumanasekera R and Vicisano L (2001) PGM Reliable Transport Protocol Specification RFC 3208 (Experimental).

Spring N, Wetherall D and Ely D (2003) Robust Explicit Congestion Notification (ECN) Signaling with Nonces RFC 3540 (Experimental).

Srikant R (2004) *The Mathematics of Internet Congestion Control (Systems and Control: Foundations and Applications)*. Springer Verlag.

Sterbenz JPG, Touch JD, Escobar J, Krishnan R and Qiao C (2001) High-speed Networking : A Systematic Approach to High-bandwidth Low-latency Communication, John Wiley.

Stevens WR (1994) *TCP/IP Illustrated, Volume 1: The Protocols*. Addison-Wesley Publishing Company.

Stevens W (1997) TCP Slow Start, Congestion Avoidance, Fast Retransmit, and Fast Recovery Algorithms RFC 2001 (Proposed Standard). Obsoleted by RFC 2581.

Stewart R, Ramalho M, Xie Q, Tuexen M and Conrad P (2004) Stream Control Transmission Protocol (SCTP) Partial Reliability Extension RFC 3758 (Proposed Standard).

Stewart RR and Xie Q (2001) *Stream Control Transmission Protocol (SCTP) - A Reference Guide*. Addison Wesley.

Stewart R, Xie Q, Morneault K, Sharp C, Schwarzbauer H, Taylor T, Rytina I, Kalla M, Zhang L and Paxson V (2000) Stream Control Transmission Protocol RFC 2960 (Proposed Standard). Updated by RFC 3309.

Su-Hsien K and Andrew LLH (1998) Performance of fuzzy logic abr rate control with large round trip times. *Proceedings of Globecom*, pp. 2464–2468, Sidney, Australia.

Su C, de Veciana G and Walrand J (2000) Explicit rate flow control for abr services in atm networks. *IEEE/ACM Transactions on Networking* **8**(3), 350–361.

Takano R, Kudoh T, Kodama Y, Matsuda M, Tezuka H and Ishikawa Y (2005) Design and evaluation of precise software pacing mechanisms for fast long-distance networks. *Protocols for Fast Long-distance Networks (PFLDnet) Workshop*, Lyon, France.

Tanenbaum AS (2003) *Computer Networks*. Pearson Education Ltd.

Tang A, Wang J and Low SH (2004) Understanding choke: throughput and spatial characteristics. *IEEE/ACM Transactions on Networking* **12**(4), 694–707.

Touch J (1997) TCP Control Block Interdependence RFC 2140 (Informational).

Tsybakov B and Georganas N (1998) Self-similar traffic and upper bounds to buffer overflow in an atm queue. *Performance Evaluation* **32**(1), 57–80.

Venkitaraman N, Mysore JP and Needham N (2002) A core-stateless utility based rate allocation framework. *Lecture Notes in Computer Science* **2334**, 1–16.

Veres A, Molnar KS and Vattay G (2000) On the propagation of long-range dependence in the internet. *SIGCOMM '00: Proceedings of the Conference on Applications, Technologies, Architectures, and Protocols for Computer Communication*, pp. 243–254. ACM Press.

Vicisano L, Crowcroft J and Rizzo L (1998) Tcp-like congestion control for layered multicast data transfer. *IEEE Infocom*, San Francisco, California, USA.

Vinnicombe G (2002) On the stability of networks operating tcp-like congestion control. In Proceedings of the IFAC World Congress, Barcelona, Spain.

Vojnovic M and Le Boudec J (2002) On the long-run behavior of equation-based rate control. *SIGCOMM '02: Proceedings of the 2002 Conference on Applications, Technologies, Architectures, and Protocols for Computer Communications*, pp. 103–116. ACM Press.

Vojnovic M, Le Boudec J and Boutremans C (2000) Global fairness of additive-increase and multiplicative-decrease with heterogeneous round-trip times. *Proceedings of the 2000 IEEE Computer and Communications Societies Conference on Computer Communications (INFOCOM-00)*, pp. 1303–1312. IEEE, Los Alamitos, CA.

Wang Z (2001) *Internet QoS: Architectures and Mechanisms for Quality of Service*. Morgan Kaufmann Publishers.

Wang Y, Claypool M and Zuo Z (2001) An empirical study of realvideo performance across the internet. *ACM SIGCOMM Internet Measurement Workshop*, San Francisco, CA.

Wang C, Li B, Hou T, Sohraby K and Lin Y (2004) Lred: a robust active queue management scheme based on packet loss ratio. *IEEE Infocom*, Hong Kong.

Wei DX, Hegde S and Low SH (2005) A burstiness control for high speed long distance tcp. *Protocols for Fast Long-distance Networks (PFLDnet) Workshop*, Lyon, France.

Welzl M (2003) *Scalable Performance Signalling and Congestion Avoidance*. Kluwer Academic Publishers.

Welzl M (2004) Tcp corruption notification options, Internet-draft draft-welzl-tcp-corruption-00, Work in Progress; Expired, Available as a Technical Report from http://www.welzl.at/publications/draft-welzl-tcp-corruption-00.txt.

Welzl M (2005a) Composing qos from congestion control elements",e+i "elektrotechnik und informationstechnik. *Elektrotechnik und Informationstechnik*.

Welzl M (2005b) Passing corrupt data across network layers: an overview of recent developments and issues. *EURASIP Journal on Applied Signal Processing* **2005**(2), 242–247.

Welzl M (2005c) Scalable router aided congestion avoidance for bulk data transfer in high speed networks. *PFLDNet 2005 Workshop*, Lyon, France.

Welzl M, Goutelle M, He E, Kettimut R, Hegde S, Gu Y, Allcock WE, Kettimuthu R, Leigh J, Primet P, Xiong C and Yousaf MM (2005) Survey of Transport Protocols Other than Standard Tcp; Global Grid Forum Draft (Work in Progress), Data Transport Research Group.

Welzl M and Mühlhäuser M (2003) Scalability and quality of service: a trade-off? *IEEE Communications Magazine* **41**(6), 32–36.

Westberg L, Császár A, Karagiannis G, Marquetant Á, Partain D, Pop O, Rexhepi V, Szabó R and Takács A (2002) Resource management in diffserv (RMD): a functionality and performance behavior overview. *Lecture Notes in Computer Science* **2334**, 17–34

Whetten B, Vicisano L, Kermode R, Handley M, Floyd S and Luby M (2001) Reliable Multicast Transport Building Blocks for One-to-many Bulk-data Transfer RFC 3048 (Informational).

Widmer J (2003) *Equation-based Congestion Control for Unicast and Multicast Data Streams*. PhD thesis, University of Mannheim.

Widmer J, Boutremans C and Le Boudec J (2004) End-to-end congestion control for tcp-friendly flows with variable packet size. *SIGCOMM Computer Communication Review* **34**(2), 137–151.

Widmer J, Denda R and Mauve M (2001) A survey on TCP-friendly congestion control. *Special Issue of the IEEE Network Magazine "Control of Best Effort Traffic"* **15**(3), 28–37.

Widmer J and Handley M (2001) Extending equation-based congestion control to multicast applications. *SIGCOMM '01: Proceedings of the 2001 Conference on Applications, Technologies, Architectures, and Protocols for Computer Communications*, pp. 275–285. ACM Press.

Widmer J and Handley M (2004) Tcp-friendly Multicast Congestion Control (tfmcc): Protocol Specification. Internet-draft Draft-ietf-rmt-bb-tfmcc-04, Work in Progress.

Xu L, Harfoush K and Rhee I (2004) Binary increase congestion control for fast long-distance networks. *INFOCOM*, Hong Kong, China.

Xue F, Liu J, Zhang L, Yang OWW and Shu Y (1999) Traffic modeling based on FARIMA models. *IEEE Canadian Conference on Electrical and Computer Engineering*, Edmonton, Alberta, Canada.

Yang YR and Lam SS (2000a) General AIMD congestion control. *IEEE ICNP*, Osaka, Japan.

Yang YR and Lam SS (2000b) Internet multicast congestion control: a survey. *ICT*, Acapulco, Mexico.

Yang YR, Kim MS and Lam SS (2001) Transient behaviors of tcp-friendly congestion control protocols. *IEEE INFOCOM 2001*, Anchorage, AK.

Yung Y and Shakkottai S (2004) Hop-by-hop congestion control over a wireless multi-hop network. *IEEE INFOCOM*, 4, pp. 2548–2558.

Zhang M, Karp B, Floyd S and Peterson L (2002) Rr-tcp: A Reordering-robust Tcp with Dsack. Technical Report TR-02-006, ICSI, Berkeley, CA.

Zixuan Z, Sung LB, ChengPeng F and Jie S (2003) Packet triplet: an enhanced packet pair probing for path capacity estimation. *Network Research Workshop*, Republic of Korea.

Index

ABC, 76, 104–106
ABE, 196
ABR, 34
ACK
    compression, 177
    congestion, 38
    starvation, 38
Active queue management, 28
Adaptive applications, 50, 52
Adaptive RED, 130–131
Admission control, 14, 189, 197–198
Aggressive framework, 203–205
AIAD, 17
AIMD, 17
ALF, 92, 102
Application autonomy, 29
Application layer framing, 92, 102
AQM, 28, 94–97, 129–137
    Adaptive RED, 130–131
    AVQ, 133–134
    BLUE, 133
    CHOKe, 135–136
    DRED, 131–132
    FRED, 135
    RED, 94–97
    RED-PD, 134
    REM, 136–137
    RIO, 137
    SFB, 133
    SRED, 132–133
    WRED, 137
ARQ, 23
    Stop-and-wait, 24
ATM
    EPD, 102

PPD, 101
ATM ABR, 34, 97–102
    CAPC, 100
    EPRCA, 100
    ERICA, 100
    OSU, 100
    RM cells, 98
    Target rate, 100
Auto-tuning, 128
Available Bit Rate, 34
AVQ, 133–134

Backwards ECN, 33
bandwidth X delay product, 23, 27
BECN, 33
BIC, 163
BIC-TCP, 163
Binomial cc, 146
BI-TCP, 163
BLUE, 133
Broadcast, 40

CADPC, 174
CADPC/PTP, 171–174
CAPC, 172
CapProbe, 169
CAVT, 18, 220–226
CAVTool, 18, 220–226
CBR, 24
CHOKe, 135–136
Choke packets, 33
Class-based queuing, 190
Client–Server, 30
Closed-loop control, 13
CM, 119–121

Network Congestion Control: Managing Internet Traffic Michael Welzl
© 2005 John Wiley & Sons, Ltd

Printed and bound by CPI Group (UK) Ltd, Croydon, CR0 4YY

27/10/2024

14580150-0004